	I B	II B	III A	IV A	V A	VI A	VII A	0
								2 **He** Helium 4.00260
			5 **B** Boron 10.81	6 **C** Carbon 12.011	7 **N** Nitrogen 14.0067	8 **O** Oxygen 15.9994	9 **F** Fluorine 18.998403	10 **Ne** Neon 20.179
			13 **Al** Aluminum 26.98154	14 **Si** Silicon 28.0855	15 **P** Phosphorus 30.97376	16 **S** Sulfur 32.06	17 **Cl** Chlorine 35.453	18 **Ar** Argon 39.948
28 **Ni** Nickel 58.69	29 **Cu** Copper 63.546	30 **Zn** Zinc 65.38	31 **Ga** Gallium 69.72	32 **Ge** Germanium 72.59	33 **As** Arsenic 74.9216	34 **Se** Selenium 78.96	35 **Br** Bromine 79.904	36 **Kr** Krypton 83.80
46 **Pd** Palladium 106.42	47 **Ag** Silver 107.868	48 **Cd** Cadmium 112.41	49 **In** Indium 114.82	50 **Sn** Tin 118.69	51 **Sb** Antimony 121.75	52 **Te** Tellurium 127.60	53 **I** Iodine 126.9045	54 **Xe** Xenon 131.29
78 **Pt** Platinum 195.08	79 **Au** Gold 196.9665	80 **Hg** Mercury 200.59	81 **Tl** Thallium 204.383	82 **Pb** Lead 207.2	83 **Bi** Bismuth 208.9804	84 **Po** Polonium (209)[a]	85 **At** Astatine (210)[a]	86 **Rn** Radon (222)[a]

metals ← → nonmetals

63 **Eu** Europium 151.96	64 **Gd** Gadolinium 157.25	65 **Tb** Terbium 158.9254	66 **Dy** Dysprosium 162.50	67 **Ho** Holmium 164.9304	68 **Er** Erbium 167.26	69 **Tm** Thulium 168.9342	70 **Yb** Ytterbium 173.04	71 **Lu** Lutetium 174.967
95 **Am** Americium (243)[a]	96 **Cm** Curium (247)[a]	97 **Bk** Berkelium (247)[a]	98 **Cf** Californium (251)[a]	99 **Es** Einsteinium (252)[a]	100 **Fm** Fermium (257)[a]	101 **Md** Mendelevium (258)[a]	102 **No** Nobelium (259)[a]	103 **Lr** Lawrencium (260)[a]

SEMIMICRO QUALITATIVE ANALYSIS

G. Brooks King
Washington State University

William E. Caldwell

Lawrence Epstein
University of Pittsburgh

Wadsworth Publishing Company · Belmont, California · A Division of Wadsworth, Inc.

Chemistry Editor: Jack Carey
Editorial Assistant: Ruth Singer
Production Editor: Andrea Cava
Managing Designer: MaryEllen Podgorski
Print Buyer: Barbara Britton
Copy Editor: Marion Hansen
Technical Illustrator: Nancy Warner
Compositor: G&S Typesetters, Inc.
Cover Photographs: Sand dunes by David Meunch;
hoar frost crystal by Stephen J. Krasemann, Peter Arnold, Inc.

Printed in the United States of America

1 2 3 4 5 6 7 8 9 10—90 89 88 87 86

ISBN 0-534-05677-6

CONTENTS

	Preface	vi
	Introduction	1
	Part One **The Semimicro Plan of Qualitative Analysis**	
1	The Semimicro Method	7
2	Schedule of Laboratory Work	11
3	Preparation of Special Equipment	12
4	Preliminary Experiments on Laboratory Techniques	15
	Part Two **Analysis for Anions**	
5	Introduction to Analysis for Anions	19
6	Action of Concentrated Sulfuric Acid on Salts	20
7	Solubility Group Tests	22
8	Specific Tests for Anions	24
	Part Three **Analysis for Cations**	
9	Introduction to Cation Analysis	35
10	Preliminary Experiment on Group Separations	37
11	Group I: The Silver Group	39
12	Group II: The Copper–Arsenic Group	43
13	Group III: The Aluminum–Nickel–Iron Group	53
14	Group IV: The Barium Group	63
15	Group V: The Alkali Group	69
	Part Four **Analysis of Alloys, Salts, and Commercial Substances**	
16	Analysis of Alloys	75
17	Analysis of Salts and Mixtures	79
18	Analysis of Commercial Substances	83
	Part Five **Problems, Exercises, and Appendix**	87
	Index	102

PREFACE

The subject of qualitative analysis, which had been de rigueur a generation ago, has languished in the years since 1968 when the last edition of Caldwell and King was published. More recently, however, chemistry teachers, awakening to the importance of descriptive chemistry in the curriculum, have returned to "qual" as an effective and enjoyable way for their students to learn practical descriptive chemistry. Most laboratory syllabi now include qual in amounts ranging from two sessions to a full semester.

This book is a modern version of the classic 1968 edition. It can serve as the laboratory text for a full-semester course; or, by selecting only certain groups of elements, it can be used for a limited number of qual sessions. Although the book has been updated, the scope, the methods, and the techniques have not been changed. The nomenclature and notation have been modernized and made consistent with the sixth edition of Mortimer's *Chemistry* and with Epstein's accompanying *Laboratory Manual for General Chemistry*. (The latter contains three brief qual experiments that are consistent with this text.) Several small changes in procedures were made to eliminate the use of reagents that are now considered potentially chronic poisons. The format of the procedure sections was changed to make them more readable, without sacrificing the facility with which one can follow the flow of precipitates and decantates from one step to the next.

Thanks are due to the Wadsworth staff for their efficient efforts in producing this book: Jack Carey, chemistry editor; Andrea Cava, production editor; MaryEllen Podgorski, designer; and Barbara Britton, print buyer.

INTRODUCTION

Safety in the Qualitative Analysis Laboratory

Semimicro quantitative analysis should be conducted with the same careful safety precautions practiced in any other chemical laboratory. Even though the hazards are mitigated by the small quantities of chemicals used, safe handling of all materials is still necessary. The greatest danger is taking the attitude that "nothing can happen." The main hazards are emphasized here. Warnings will be repeated where relevant in the procedures.

Eye protection

Students must wear eye protection continuously in the qualitative analysis laboratory because the eyes are vulnerable to damage by even the slightest drop of chemical reagent. Students who wear glasses must wear safety goggles over the glasses. Depending on state laws and specific school rules, glasses with side shields or special safety glasses with side shields may be adequate.

Students may not wear contact lenses in the chemical laboratory. Wear regular glasses plus safety goggles, or safety glasses with side shields if permitted.

Fire hazards

Every student will probably have a lighted flame, usually to keep a water bath simmering. In several procedures volatile flammable organic liquids, such as ethyl alcohol, are used. To minimize the fire hazard, students should obtain just the required quantity of these liquids from the reagent shelf when needed (NOT sooner), and use it right away. NEVER carry the reagent bottle to your desk.

Acute poisons

Various procedures require the use of potassium cyanide (KCN) and other cyanides; hydrogen sulfide (H_2S) or its precursor, thioacetamide; and hydrofluoric acid (HF) or fluoride salts. Even in the small quantities required in semimicro work, cyanides are deadly poisonous if ingested. If a strong acid is added to a cyanide salt, deadly fumes of hydrogen cyanide (HCN) are evolved.

Invariably qualitative analysis laboratories are plagued by the "rotten-egg" smell of

H_2S gas because it is used so often in cation analysis. If possible, students should perform work in which H_2S is evolved under a fume hood, although there will rarely be sufficient space for all. The laboratory must be forcibly ventilated. If the stench becomes very unpleasant, the instructor should stop all work and evacuate the laboratory. H_2S fumes are more deadly than those of HCN, but fortunately the odor of H_2S is so unpleasant that low concentrations in the air are easily detected. Students should not think they are being brave because they endured the smell of H_2S, while their more timid classmates had to leave the laboratory.

HF is very dangerous if ingested or inhaled. But, in addition, it is also very dangerous if contacted by the skin. Handle this acid carefully and rinse copiously with water if it is accidentally spilled on the skin. Remember that HF is a weak acid, and so it will form whenever a strong acid is added to any fluoride solution.

Chronic poisons

Many chemicals traditionally used in instructional laboratories have been identified recently as chronic poisons. Many appear on lists of suspected carcinogens because they have shown mutagenic effects in broad screening tests. Remember that the danger arises from constant daily exposure over periods of years. If students work reasonably neatly and use only the recommended amounts of chemicals, the danger of overexposure to these chronic poisons is very, very slight.

In this edition of *Semimicro Qualitative Analysis*, any confirmed chronic poisons have been eliminated. We do call to your attention the few listed below and suggest that you rinse your hands promptly and thoroughly after handling them.

1. Chromium salts, especially Cr(VI)
2. Halogenated organic solvents
3. Thioacetamide

Some instructors may wish to eliminate use of thioacetamide, but it is by far the most convenient source of H_2S in small amounts. The alternatives may be worse because of the lack of control and consequent contamination of the laboratory with H_2S fumes. Alternative sources of $H_2S(g)$ are:

1. Steel tanks of compressed H_2S
2. Laboratory generators in which FeS is treated with H_2SO_4
3. Laboratory generators in which a mixture of sulfur and petroleum wax is heated

Other hazards

For practically all students this is not the first laboratory course, and so we will not repeat the usual safety instructions concerning handling hot liquids, working with glass and rubber, and so on. Refer to your regular laboratory manual for additional safety information.

The centrifuge is used frequently in qualitative analysis. If the centrifuge is not properly balanced when loaded with tubes of solution, it becomes a mechanical hazard; it will vibrate severely and "walk" across the bench top. Be careful not to touch the rotor when the centrifuge is running; it can injure your fingers and hands. After turning off the centrifuge, allow it to coast to a stop and do not use your hands as a brake.

An Introduction to Qualitative Analysis

Qualitative analysis is that branch of analytical chemistry that determines the elements or ions present in an unknown substance or mixture of substances. Quantitative analysis is concerned with the amount of an element or compound present. For example, a qualitative analysis may indicate the presence of lead, sulfur, silica, and arsenic; a quantitative analysis may show 20% lead, 25% sulfur, 2% arsenic, and 53% silica.

This manual will consider only *qualitative analysis*. One can sometimes distinguish qualitatively between two substances by some difference in the way they affect the senses—taste, sight, or smell. For example, salt and sugar differ in taste; carbon and sulfur differ in color and crystalline form. At times it is not so easy to detect differences, and then substances must be more closely examined for their chemical or physical properties. It is possible to differentiate the colorless salts $AgNO_3$ and $NaNO_3$ by such physical properties as melting point, crystalline form, or degree of solubility in water, for example. A more precise means of differentiation, however, is the addition of HCl to their water solutions; a white, curdy precipitate (AgCl) is formed with one, and no precipitate with the other.

There is greater difficulty in identifying the various constituents of a mixture. The comparative solubility of different substances is the principal basis for their separation. The difference might be in *physical solubility*, such as NaCl in H_2O or I_2 in CCl_4, or in *chemical solubility*, such as $CaCO_3$ in HCl or AgCl in NH_3.

The usual procedure for testing an unknown is dissolving the sample and testing the resulting solution for the ions that may be present. An alloy, for example, when acid dissolved gives a solution of the metals as ions (cations). Dissolving of an acid, base, or salt gives respective cations and anions of the elements or the complex ions present. Thus, qualitative analysis is commonly divided into *anion analysis* and *cation analysis*. Test procedures involve a separation of the ions by precipitation and solution, followed by specific tests for each ion. The specific tests may give characteristically colored solutions or precipitates on the addition of certain reagents. For example, the addition of $K_4Fe(CN)_6$ (potassium ferrocyanide) to a solution containing Fe^{3+} gives a deep blue precipitate of $KFe[Fe(CN)_6]$ called Prussian blue.

An analysis performed on a solution of the unknown sample may not always indicate how the ions were originally combined. Obviously, if only Na^+ and Cl^- are found, the original substance must have been NaCl. If Na^+, K^+, Cl^-, and NO_3^- are found, the original may have been a mixture of NaCl and KNO_3, or of $NaNO_3$ and KCl, or all four salts. To identify the substances present in such a mixture, additional tests must be made; these are usually of a physical nature, such as examination of crystal structure, color, and so on.

Many industrial-process separations are based on principles used in qualitative analysis. One specific example is the purification of aluminum ore, which may be considered a mixture of aluminum hydroxide and iron hydroxide. NaOH will dissolve $Al(OH)_3$ and not $Fe(OH)_3$ (see similar chemical treatment in an analytical scheme to follow). The basic chemistry of both nonmetals and metals is encountered in the procedures of analytical chemistry.

This manual has been revised to conform to modern practices with the exception that the chemical equations used to illustrate the procedures retain two old-fashioned but useful devices:

1. Formulas for substances that precipitate from solution are underlined.
2. Evolution of a gas is indicated by an arrow pointing upward.

Many of the general principles and concepts learned in elementary chemistry are applicable to qualitative analysis. Since such principles are covered in the text, those topics that are particularly pertinent to qualitative procedures are listed below along with their location.

- Manner of expressing amounts of a solute, Chap. 3: 3.3, 3.4
- Methods of expressing the concentration of a solution, Chap. 4: 4.5, 4.6; Chap. 12: 12.6
- Law of chemical equilibrium, Chap. 15: 15.1–15.4
- Ionic nature of acids, bases, and salts in water solution, Chap. 12: 12.11
- Exponential numbers, Appendix C
- Ionic equilibrium, Chap. 17: 17.1–17.7
- Solubility product principle, Chap. 18: 18.1–18.3
- Oxidation number, Chap. 13: 13.2
- Chemical changes in qualitative analysis
  - Ionic union, Chap. 13: 13.1
  - Oxidation-reduction, Chap. 13: 13.2, 13.3
  - Electrode potentials, Chap. 20: 20.5, 20.6, 20.7
  - Formation of complex salts and complex ions, Chap. 18: 18.4
  - Amphoterism, Chap. 18: 18.5
  - Hydrolysis, Chap. 17: 17.8

PART ONE

THE SEMIMICRO PLAN OF QUALITATIVE ANALYSIS

THE SEMIMICRO METHOD

1

Macro methods of qualitative analysis involve the use of rather large quantities of materials; in the case of solids, amounts of 1 g or more are often used; in the case of liquids, volumes of 100 mL or more are common. The semimicro method is an adaptation of the macro method of qualitative analysis to very much smaller quantities of substances—that is, milligram amounts of solids and drop or milliliter quantities of liquids.

Although tests in semimicro procedures are not more sensitive than those of the macro methods, the former have a great advantage in the saving of time and chemicals. Whereas filtrations, washing of precipitates, and evaporations may require up to an hour on the macro scale, these operations can be carried out in a few minutes on the semimicro basis. As a result, the student can analyze more samples in the time available and thus gain a much wider training in analytical procedures.

Terms Used in Qualitative Analysis

Centrifuge is a device used to accelerate the settling of precipitates by taking advantage of centrifugal force.

Colloidal precipitates are insoluble particles that have a size between those in suspension and those in true solution. They show Brownian movement and Tyndall effect, and do not readily settle from solution.

Decantation is pouring liquid from a settled precipitate.

Filtration is the separation of a precipitate from a liquid by passage through a porous medium. The *filtrate* is the solution passing through the filter. The *residue* or precipitate remains on the filter paper. (In semimicro analysis, separations are usually performed by centrifuging and decantation.)

Ignition is the strong heating of solids.

Precipitation is the formation of an insoluble solid by the reaction of a reagent on a solution. The insoluble solid is called the *precipitate*.

Reagent is a substance used to produce a chemical change in the material under consideration. For example, HCl is a reagent when added to a solution of $AgNO_3$ because it produces a precipitate of AgCl.

Turbidity is a cloudy condition due to suspended matter in a solvent.

Washing is the cleansing of a precipitate (usually by H_2O) to remove the adhering solution from which it has been precipitated.

Semimicro Equipment

In general, items of semimicro equipment are much smaller and simpler than the conventional pieces of macro equipment. (A recommended list of items of equipment is given in the Appendix.) One of the features of the semimicro method is the use of a centrifuge (Figure 1) in separating precipitates from solutions. Centrifuges, which should be conveniently located near the working spaces in the laboratory, may be shared by several students.

There are various ways in which chemicals may be provided and arranged for semimicro work. Each laboratory bench may be equipped with a set of chemicals to be shared by several students. Each student may have a small set of frequently used chemicals, such as the common acids and bases. Several students could then share a larger set of less commonly used reagents. In general, liquids and solutions should be made available in dropping bottles of 15 to 100 mL capacity. Solids may be stored in widemouth salt bottles. An arrangement which has been used successfully is described below.

Each student outfit should include a block of liquid reagents which are used frequently—for example, dilute and concentrated HCl, dilute and concentrated H_2SO_4,

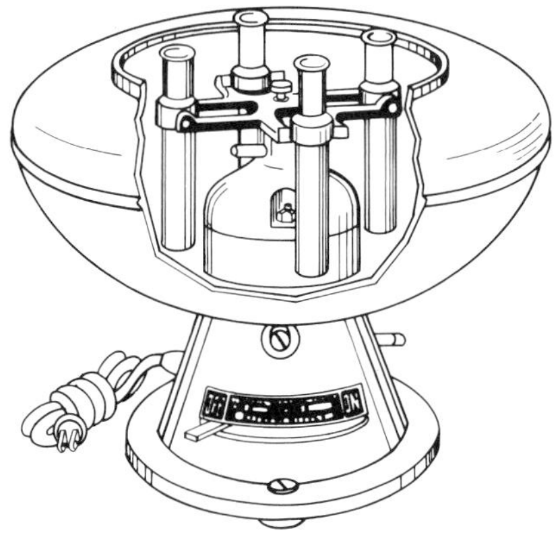

Figure 1

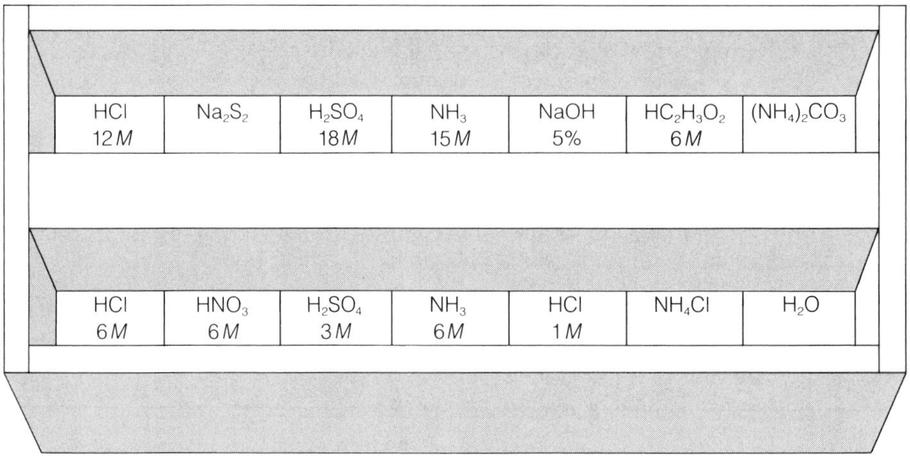

Figure 2

and so on. The dropping bottles may be filled from side shelf bottles as the need arises. These individual reagent sets consist of 14 square dropping bottles (15 mL) with screw caps (Figure 2). Less frequently used liquid and solid reagents are stored in larger bottles (about 100 mL) placed on side shelves. Bottles for liquid reagents should be equipped with medicine droppers and bulbs; bottles of solids may be equipped with spatulas inserted through the stoppers.

Laboratory Techniques

Probably the most important technique to develop in semimicro analysis is the proper handling of precipitates. Precipitations are usually carried out in small (3- or 4-inch) test tubes by the drop-by-drop addition of reagents from medicine droppers. The tube is then centrifuged to settle the precipitate. Complete precipitation is essential and should be tested for by the addition of a drop of the precipitating reagent to the supernatant liquid above the precipitate after centrifuging. If no further precipitation takes place in the supernatant liquid, precipitation may be assumed complete.

The centrifuge must be balanced by placing a tube of the same size containing an amount of water equal to the weight of the test solution opposite to the tube containing the test solution. If the centrifuge vibrates, it should be turned off and the trouble investigated. The time of centrifuging depends on the nature of the precipitate, but 15–20 seconds will usually suffice. As a result of the centrifuging, the solid becomes packed tightly in the end of the tube. The supernatant liquid (sometimes referred to as *decantate*) may be removed by decanting without disturbing the precipitate. Alternatively, the supernatant liquid may be drawn up into a medicine dropper with a long thin tip (a Pasteur pipet).

After the precipitate and solution are separated, the precipitate is washed by adding a few drops of the washing reagent (usually water), mixing thoroughly with a stirring rod, centrifuging, and removing the washings with the aid of a medicine dropper tube or by decantation. Two or more washings are usually necessary. Failure to wash precipitates thoroughly is one of the most common sources of error in qualitative analysis.

Evaporations are usually carried out in a small evaporating dish over a beaker of boiling water, which serves as a hot water bath, or by cautious heating of a solution in a crucible with a low flame to prevent spattering.

Absolute cleanliness is essential to success in semimicro analysis. Great care should be taken to prevent contamination of liquid or solid reagents. It is good practice to cover the working space on the laboratory table with a clean towel. A beaker of clean water reserved for clean droppers, stirring rods, and spatulas is also helpful.

Solubility Rules*

Since solubility of substances in water is an important consideration in analysis, the student should be familiar with the following solubility rules.

---

*These brief solubility rules are not all-inclusive; the consideration of less common compounds would call for marked augmentation of the rules. For example, this first solubility rule needs additional notes such as: Tl(I), Au(I), and Cu(I) halides are insoluble. Salts such as $SnCl_4$, $BiCl_3$, or $SbCl_3$ hydrolyze so markedly when put in water that acid is needed to keep them in solution; for example,

$$BiCl_3 + H_2O \rightleftharpoons BiOCl\downarrow + 2\,HCl$$

1. All chlorides, bromides, and iodides are soluble in water, except those of lead, silver, and mercury(I); mercury(II) iodide is also insoluble in water. Lead chloride is quite soluble in hot water.

2. All nitrates, nitrites, chlorates, and acetates are soluble in water.

3. All potassium, sodium, and ammonium salts are soluble in water, except K_2PtCl_6, $(NH_4)_2PtCl_6$, $KHC_4H_4O_6$, $K_3Co(NO_2)_6$, Na_2SiF_6, and $NaSb(OH)_6$, which are sparingly soluble.

4. All oxides and hydroxides are insoluble in water except those of the alkali metals and those of barium, strontium, and calcium, which are sparingly soluble.

5. All sulfides except ammonium sulfide and those of the alkali and alkaline earth metals are insoluble in water.

6. The sulfates of barium, strontium, and lead are insoluble in water and acids. The sulfate of calcium is slightly soluble. All other sulfates are soluble.

7. All carbonates, phosphates, borates, oxalates, chromates, arsenites, arsenates, ferrocyanides, and ferricyanides except those of the alkali metals are insoluble or only very slightly soluble in water ($CaCrO_4$ and $MgCrO_4$ are fairly soluble). All are readily soluble in dilute reagent acids.

8. All silicates are insoluble in water except those of the alkali metals.

9. All cyanides are insoluble in water except those of the alkali metals and mercury(II).

SCHEDULE OF LABORATORY WORK

2

A suggested schedule of laboratory work for a course in semimicro qualitative analysis is outlined below.

A. Preparation of special equipment
B. Preliminary experiment on laboratory techniques
C. Tests for anions
    1. Sulfuric acid treatment
    2. Group tests
    3. Specific tests
D. Analysis of anion unknowns
E. Preliminary experiment on group separations
F. Analysis for cations by groups
    1. Group I. The silver group
        a. Preliminary tests
        b. Analysis of known solution
        c. Analysis of unknown solution
    2. Group II. The copper–arsenic group
        a. Preliminary tests
        b. Analysis of known solution
        c. Analysis of unknown solution
    3. Group III. The aluminum–iron–nickel group
        a. Preliminary tests
        b. Analysis of known solution
        c. Analysis of unknown solution
    4. Group IV. The barium group
        a. Preliminary tests
        b. Analysis of known solution
    5. Group V. Alkali group
        a. Analysis of a known solution
        b. Analysis of unknown solution for Groups IV–V
G. Analysis of general unknown (all cations)
H. Analysis of unknown alloy
I. Analysis of single salts for cation and anion
J. Analysis of salt mixtures for cations and anions

3 PREPARATION OF SPECIAL EQUIPMENT

Wash Bottle

Use a plastic wash bottle (also called a squeeze bottle) obtainable from most chemical supply houses (see Figure 3).

Stirring Rods and Spatulas

Prepare four stirring rods by cutting two 10-inch lengths of 3-mm solid glass rod. Heat each in the middle until it is soft, remove from the flame, and draw it out to about 1 mm diameter. Cut it in two pieces, and reheat the small end until a small ball forms. Heat the opposite end until a larger ball forms and, while it is hot, press it down firmly on a heat resistant surface, forming a head or plunger (used for mixing solutions in small test tubes). (See Figure 4.)

Figure 3

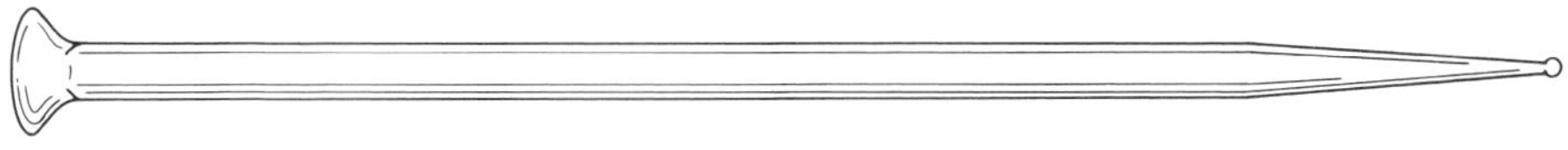

Figure 4

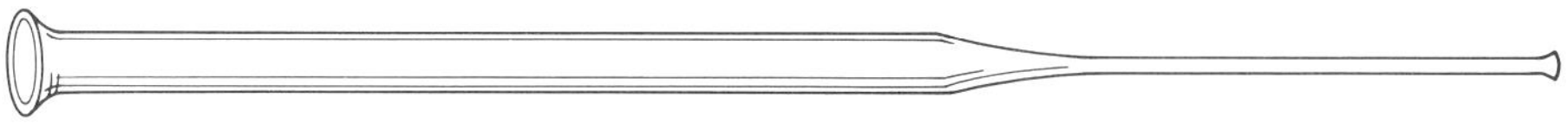

Figure 5

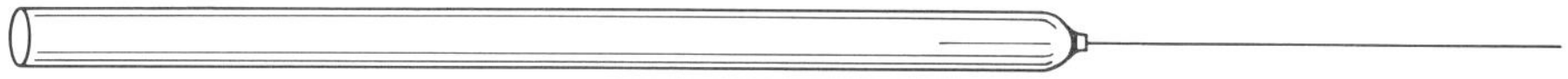

Figure 6

Prepare four spatulas from 5-inch lengths of glass rod. Heat one end of the rod until a bead forms. Remove it from the flame and pinch it flat with tongs or between other metal surfaces. Avoid getting the widened portion too wide to reach the bottom of small test tubes. Put a knob on the other end of the rod.

Pipets

Prepare two pipets for drawing off the liquid above a precipitate (supernatant liquid) in a small test tube after centrifuging. The pipets are like medicine droppers except that the small end should be drawn out about 2 inches for easier insertion into small-diameter test tubes (Figure 5). Such pipets, called Pasteur pipets, are commercially available.

Platinum or Nichrome Wire for Flame Tests

Platinum wire, traditionally used for flame tests, has become so costly that nichrome wire is often used instead. If platinum wire is available, seal a 2-inch length of wire into one end of a piece of soft glass tubing (Figure 6).

Alternatively, a short length of nichrome wire can be used. It need not be sealed in glass as long as it is held at a point at least 4 inches away from the flame. Consider the nichrome wires expendable and discard them after a term or two of use.

Flame test

In a flame test, the substance to be tested should be a solid or a concentrated solution. If a solution is to be tested, evaporate to dryness and moisten the residue with a drop of concentrated HCl. Heat the platinum (or nichrome) wire to redness and touch it to the solid in such a way that some solid adheres to the wire. Introduce the wire again into the flame and observe the color of the flame produced. Make flame tests on solid NaCl, $BaCl_2$, $SrCl_2$, and $CaCl_2$. Before each test clean the wire by dipping it into concentrated HCl in a watch glass and reheating in the flame until no color is evident. Record the colors produced.

Preparation of a Borax Bead

Make a circular loop about 2–3 mm in diameter at the end of the platinum or nichrome wire prepared above. Heat the wire to redness in the flame and quickly plunge it into powdered borax. Heat in the flame until the borax fuses to a clear colorless bead.

Borax bead test

In making a borax bead test, heat the bead until it just softens, then touch it to a small bit (a bulk of about 1/10 that of the head of a pin) of the substance to be tested. Heat again in the oxidizing flame (above the inner blue cone of the flame). Allow to cool and observe the color. It is important that not too much substance be used in the test, since an excess may give the bead a black and opaque appearance. Make a bead test on samples of $Co(NO_3)_2$, $Ni(NO_3)_2$, and $FeCl_3$. Record the colors produced.

PRELIMINARY EXPERIMENTS ON LABORATORY TECHNIQUES

4

Laboratory Technique

The purposes of this experiment are (1) to become familiar with the use and care of the centrifuge, (2) to learn the technique of handling and washing small precipitates, and (3) to emphasize the value of careful techniques.

Place 1 drop of Hg_2^{2+} test solution and 1 drop of Ba^{2+} test solution in a test tube. Add 10 drops of water and stir the solution. Add 2 drops dilute HCl (1 *M*) and note the precipitation of mercury(I) chloride. Mix by shaking the tube or, better, by stirring with the "plunger" stirring rod prepared earlier. Centrifuge the mixture after carefully balancing it with another tube containing water. Fifteen to twenty seconds of centrifuging is long enough to settle most precipitates. Carefully add another drop of dilute HCl to the supernatant liquid to test for complete precipitation.

Draw off the supernatant liquid (decantate or filtrate) with one of the pipets you have prepared. Place the solution in a test tube, neutralize with dilute NH_3, and add a drop of K_2CrO_4 solution. The yellow precipitate is $BaCrO_4$, and shows that barium ion remains in solution after centrifuging. Wash the residue of mercury(I) chloride obtained above by stirring with 10 drops of water. Centrifuge and test the washings again by addition of K_2CrO_4 to see if any barium ion remains. Repeat the washings until barium ion is no longer found. How many washings were necessary to remove Ba^{2+}?

Write ionic equations for the reactions taking place in the above procedure.

Use of Thioacetamide as a Source of H_2S

The formation of insoluble sulfides is the basis of a considerable portion of cation analysis. The source of sulfide ions for the precipitations is H_2S, which is most conveniently produced from thioacetamide. This substance hydrolyzes quite readily in water at 70°–90°C, but remains relatively stable in aqueous solution at room temperature.

$$CH_3CSNH_2 + 2\ H_2O \rightarrow CH_3COONH_4 + H_2S$$

An 8% solution of thioacetamide may be bottled as a laboratory reagent. When H_2S is needed in semimicro qualitative precipitations, satisfactory results are obtained by adding 6 drops of the 8% thioacetamide to the cation solution in a test tube, which is then placed in a boiling water bath for 5 minutes (repeated once or twice if necessary). In adding the thioacetamide solution, use a glass stirring rod to ensure mixing. The ammonium acetate formed in the reaction has some merit in aiding the coagulation of the precipitate.

Place 1 mL of test solutions of Cu^{2+}, As^{3+}, Sb^{3+}, Zn^{2+}, Mn^{2+}, Ca^{2+}, and K^{+}, respectively, in small test tubes. Add 1 mL of water and 5 drops of 1 *M* HCl to each tube. The solution in each tube is now to be saturated with H_2S. Add 6 drops thioacetamide solution to each tube, and place the tubes in a boiling water bath as described above.

To the test tubes containing cations where no precipitation took place, add dilute NH_3 dropwise to basicity. What two cations now precipitate as sulfides in basic solution? Record the colors of all the precipitated sulfides. What two cations of those tested do not form water-insoluble sulfides?

A preliminary observation from this experiment is that certain cations will precipitate as sulfides in the presence of acid, some precipitate as sulfides only in neutral or basic solution, and others do not precipitate from water solution as sulfides. This generalization is used in chemically testing an unknown cation solution. A cation unknown can at least be narrowed down as belonging to one of three groups according to its action with H_2S (thioacetamide).

PART TWO

ANALYSIS FOR ANIONS

INTRODUCTION TO ANALYSIS FOR ANIONS

5

Negatively charged ions, whether monatomic, like the halide ions, or polyatomic, such as the sulfate and nitrate ions, are called *anions*, because they migrate to the anode or positive pole during electrolysis. Similarly, positively charged ions are called *cations* because they migrate toward the cathode or negative pole during electrolysis. The analysis of a sample divides itself into two sections virtually independent of each other; one is concerned with the identification of cations, the other with the identification of anions. In this manual, anion analysis will be considered first because the tests for anions are relatively simple and may be carried out with a minimum of interference from other ions present. Although a systematic procedure for separation and identification of anions may be developed, such a procedure is tedious and time-consuming. In general, anions are adaptable to identification tests that are quickly and simply carried out.

After one becomes thoroughly familiar with procedures for testing for both cations and anions, it is usually advantageous to test for cations first, since, as will be pointed out later, elimination of certain anions may be indicated on the basis of the cation analysis.

This manual will describe tests for the following anions:

Sulfate, SO_4^{2-}	Arsenate, AsO_4^{3-}	Iodide, I^-
Sulfite, SO_3^{2-}	Phosphate, PO_4^{3-}	Sulfide, S^{2-}
Carbonate, CO_3^{2-}	Chloride, Cl^-	Nitrate, NO_3^-
Borate, BO_3^{3-}	Bromide, Br^-	Nitrite, NO_2^-
Chromate, CrO_4^{2-}		Acetate, $C_2H_3O_2^-$

It is customary to apply the following indicative tests in the course of analysis for anions: (1) action of concentrated H_2SO_4 on the sample (if a solid), (2) solubility group tests on a solution of the sample, and (3) specific or individual tests.

To familiarize students with these procedures, sodium salts of the anions listed above are available in the laboratory for the tests (except that ZnS or CaS is preferable to sodium sulfide). Before you proceed to the analysis of the unknown, apply the three tests above to each of the known sodium salts. (If desired, the class may be provided with a set of test solutions of the sodium salts of the anions for making group tests and certain of the specific tests.) Follow Procedures 1, 2, and 3 in making the above three tests. (See pages 20, 22, and 28.)

6 ACTION OF CONCENTRATED SULFURIC ACID ON SALTS

A great deal of information can be obtained concerning the character of the anion of a compound by heating it moderately with a high-boiling acid (concentrated H_2SO_4 or H_3PO_4), which produces the acid of the unknown anion. This acid may decompose or be oxidized, and a gas may be evolved that can be identified by color or odor. A small amount of solid material (0.1 g or less) should be used for this test. Avoid breathing any unknown gas deeply, and never point the heated tube at your neighbor.

Hot concentrated H_2SO_4 is very corrosive, and CAUTION should be exercised in the heating outlined in the procedure below. Follow carefully the directions for the amounts to be used. Raise the temperature *gradually* while looking for evidence of chemical change.

Apply Procedure 1 to each of the sodium salts of the anions listed on page 19.*

Procedure 1 Action of Sulfuric Acid on Sample

Place about 0.1 g of solid salt (about as much as can be heaped on ⅛ inch of the end of the spatula) in a small test tube and moisten with a few drops of concentrated H_2SO_4. Warm gently. DO NOT BOIL! Indications of possible anions present are summarized in the table below.

Result of test	*Inference as to possible anion*
No change	Sulfate, Phosphate, Borate, Arsenate
Sharp odor of vinegar or acetic acid	Acetate
Frothing and sharp odor of SO_2, effervesces Solid sulfite also emits SO_2 with cold dilute HCl	Sulfite
Odor of rotten eggs; H_2S, S, and possibly SO_2	Sulfide (Iodide)
Brownish vapor, characteristic odor and color of bromine	Bromide (Nitrite)

(continued)

*Instructors may choose to delete testing for certain anions, such as arsenate, nitrite, and chromate.

Procedure 1 *(continued)*

Result of test	*Inference as to possible anion*
Black solid, violet vapor, odor of H_2S	Iodide
Brown vapor of NO_2 (in cold), vigorous effervescence	Nitrite (Bromide)
Nitric acid, vaporizing on *high* heating One may notice a little brown NO_2 against a white background if some bit of reducing impurity is present. Otherwise, the HNO_3 colorless fumes condense on the cooler upper part of the test tube. After some Cu shreds have been placed in the mouth of the tube and the tube heated, the HNO_3, as it evolves, gives characteristic action on Cu with brown NO_2 formed, and some green $Cu(NO_3)_2$.	Nitrate
Orange to red color from yellow	Chromate
Effervescence, sharp odor of HCl	Chloride
Vigorous effervescence, colorless, odorless gas	Carbonate

Write balanced equations for the reaction of sulfuric acid on these salts.

7 SOLUBILITY GROUP TESTS

The anions are grouped as follows:

Group I	The sulfate group: SO_4^{2-}, CO_3^{2-}, CrO_4^{2-}, AsO_4^{3-}, PO_4^{3-}, SO_3^{2-}, BO_3^{3-}
Group II	Halide group: Cl^-, Br^-, I^-, S^{2-}
Group III	Nitrate group: NO_3^-, NO_2^-, $C_2H_3O_2^-$

This division of the anions is based on the following facts: (1) In ammoniacal solutions only the anions of Group I are precipitated with Ba^{2+}; (2) in HNO_3 solution only the anions of Group II are precipitated with Ag^+. Accordingly, the group reagent for Group I is NH_3 and $BaCl_2$, the group reagent for Group II is HNO_3 and $AgNO_3$, and there is no group reagent for Group III, since all nitrates, nitrites, and acetates are soluble.

Prepare (or obtain from side shelf if test solutions of anions have been provided) 1 mL aqueous solutions of the sodium salts of all the anions listed above. (Dissolve a few crystals, a volume about that of the head of a safety match, of the salt in 1 mL H_2O. The solutions used in making the group tests and specific tests must not be too concentrated, since some of the tests may be invalidated in concentrated solution.) Apply Procedure 2 to each solution and record whether a precipitate forms and its color. Arrange your results as suggested in the table on the following page. Enter the color of the precipitate for each combination that gave a positive result.

Procedure 2 Group Tests for Anions*

Sulfate Group: To 5 drops of prepared solution add dilute NH_3 (1) until basic, then 2 drops $BaCl_2$ solution. Formation of precipitate indicates anions of this group. If no precipitate is formed immediately, allow to stand for a few minutes, since barium borate forms slowly (2), (3). Now add concentrated HCl drop by drop until the solution is acid. If precipitate dissolves completely, SO_4^{2-} is absent; otherwise, SO_4^{2-} is present.

Halide Group: To 5 drops of prepared solution, add 5 drops dilute HNO_3 (4). Heat, and then add 1 drop $AgNO_3$ solution. Formation of precipitate indicates

(continued)

*Numbers in parentheses refer to Notes (below).

Procedure 2 *(continued)*

anions of this group. (If a sulfite is present, silver sulfite may precipitate; if so, it is decomposed to SO_2 by heating with HNO_3.)

Nitrate Group: If no precipitate is obtained above, anions of this group are probably present.

Notes

1. The solution must be neutral or basic, since the barium salts of most of the sulfate group are soluble in acid solution and would not precipitate.

2. The absence of a precipitate does not definitely eliminate borate. Always make a specific test for borate.

3. Since BaS is slightly soluble in water, a cloudiness may result as Ba^{2+} is added to a sulfide.

4. If the solution is not made acid with HNO_3, some Group I anions (if present) would precipitate as silver salts.

Anion	Sulfate group (Insoluble barium salt group)	Halide group (Insoluble silver salt group)	Nitrate group (Soluble group)
Sulfate			
Borate			
Phosphate			
Chloride			
Bromide			
Sulfite			
Arsenate			
Carbonate			
Sulfide			
Iodide			
Nitrite			
Nitrate			
Acetate			
Chromate			

8 SPECIFIC TESTS FOR ANIONS

Following the preliminary tests on a sample for anions (that is, H_2SO_4 treatment if the sample is a solid and group tests according to Procedure 2), specific or individual tests are made for those anions that have not been eliminated as possibilities. The chemistry involved and the basis for these specific tests are summarized below. See Procedure 3 for specific directions.

Sulfate

Since $BaSO_4$ is the only barium salt (of the anions considered here) insoluble in HCl, the addition of Ba^{2+} ($BaCl_2$) and HCl to a solution yields a precipitate of $BaSO_4$ if SO_4^{2-} is present.

$$Ba^{2+} + SO_4^{2-} \rightarrow \underset{\text{white}}{\underline{BaSO_4}}$$

Arsenate

If AsO_4^{3-} ion is present, arsenic should be found in the cation analysis, since H_2S reacts with this ion to give As_2S_5.

$$2\ AsO_4^{3-} + 6\ H^+ + 5\ H_2S \rightarrow \underline{As_2S_5} + 8\ H_2O$$

The specific test for AsO_4^{3-} depends on the precipitation of white crystalline $MgNH_4AsO_4$ from a solution containing AsO_4^{3-} with magnesia mixture ($MgCl_2$, NH_4Cl, NH_3).

$$Mg^{2+} + NH_4^+ + AsO_4^{3-} \rightarrow \underset{\text{white}}{\underline{MgNH_4AsO_4}}$$

Since PO_4^{3-} gives a similar precipitate ($MgNH_4PO_4$) with magnesia mixture, AsO_4^{3-} is confirmed by treatment of the precipitate (either $MgNH_4AsO_4$ or $MgNH_4PO_4$) with acetic acid and silver nitrate solution. If arsenate ion is present, a chocolate-colored precipitate of silver arsenate is obtained.

$$AsO_4^{3-} + 3\,Ag^+ \rightarrow \underset{\text{chocolate}}{\underline{Ag_3AsO_4}}$$

If phosphate ion is present, a yellow precipitate of Ag_3PO_4 is obtained.

$$PO_4^{3-} + 3\,Ag^+ \rightarrow \underset{\text{yellow}}{\underline{Ag_3PO_4}}$$

Phosphate

The addition of ammonium molybdate and HNO_3 to a solution containing phosphate ion yields a finely divided canary-yellow precipitate of ammonium phosphomolybdate.

$$PO_4^{3-} + 12\,(NH_4)_2MoO_4 + 21\,HNO_3 + 3\,H^+ \rightarrow$$
$$\underset{\text{yellow}}{\underline{(NH_4)_3PO_4 \cdot 12\,MoO_3}} + 21\,NH_4NO_3 + 12\,H_2O$$

Since arsenate ion gives a similarly colored yellow precipitate of ammonium arseno-molybdate, any arsenic must be removed before the test. This can be accomplished by its precipitation as As_2S_5 with H_2S.

Borate

If concentrated H_2SO_4 and methyl alcohol (or ethyl alcohol) are added to a borate, methyl borate (or ethyl borate) is produced. When burned this gives an immediate greenish colored flame.

$$Na_2B_4O_7 + H_2SO_4 + 5\,H_2O \rightarrow Na_2SO_4 + 4\,H_3BO_3$$
$$H_3BO_3 + 3\,CH_3OH \rightarrow (CH_3)_3BO_3$$
$$2\,(CH_3)_3BO_3 + 9\,O_2 \rightarrow 6\,CO_2 + B_2O_3 + 9\,H_2O$$

CAUTION: Avoid breathing fumes of methyl borate or methyl alcohol.

Chromate

If this ion is present, chromium will be found in the cation analysis. Chromates are yellow, and this often aids in their identification. The specific test for the ion depends on the precipitation of $BaCrO_4$ in acetic acid solution.

$$Ba^{2+} + CrO_4^{2-} \rightarrow \underset{\text{yellow}}{\underline{BaCrO_4}}$$

Chromates in acid solution turn to an orange-colored dichromate.

$$2\,CrO_4^{2-} + 2\,H^+ \rightarrow H_2O + Cr_2O_7^{2-}$$

If a few drops of H_2O_2 and ether are shaken into a dilute, slightly acid chromate solution, a blue color develops in the ether layer. This is probably because of the formation of a peroxide CrO_5. This compound is not stable and the blue color will fade. All solutions must be cold.

Sulfite

The test for sulfite ion is based on its oxidation to SO_4^{2-} with H_2O_2, followed by precipitation of $BaSO_4$ in HCl solution.

$$SO_3^{2-} + H_2O_2 \rightarrow SO_4^{2-} + H_2O$$

$$SO_4^{2-} + Ba^{2+} \rightarrow \underline{BaSO_4}$$

If sulfate ion is present, it must be removed by precipitation as $BaSO_4$ before H_2O_2 is added. Furthermore, a solid sulfite will emit SO_2 when treated with dilute HCl.

Carbonate

Carbonates effervesce rapidly on the addition of HCl to give colorless, odorless CO_2. Effervescence in itself is a good indication of the presence of a carbonate, but, since sulfides and sulfites also effervesce with HCl, this is not a specific test. To confirm the carbonate ion, the gas evolved on addition of HCl is allowed to react with $Ba(OH)_2$ or $Ca(OH)_2$ solution, which yields a white precipitate of $BaCO_3$ or $CaCO_3$.

$$CO_3^{2-} + 2\ H^+ \rightarrow CO_2 + H_2O$$

$$CO_2 + Ba(OH)_2 \rightarrow \underline{BaCO_3} + H_2O$$

Since SO_2 may give a white precipitate of $BaSO_3$ with barium hydroxide, sulfite ion must be removed if present before testing for carbonate. This can be accomplished by oxidation of sulfite ion to sulfate ion with sodium peroxide (or hydrogen peroxide).

$$2\ SO_3^{2-} + 2\ Na_2O_2 + 2\ H_2O \rightarrow 2\ SO_4^{2-} + 4\ Na^+ + 4\ OH^-$$

Sulfide

Sulfide ion is identified by liberating H_2S, which will react with a solution of lead acetate or lead nitrate to produce a black precipitate of PbS.

$$Pb^{2+} + H_2S \rightarrow \underline{PbS} + 2\ H^+$$

Chloride, Bromide, and Iodide

These ions form insoluble silver halide salts with a solution of silver nitrate and nitric acid. Although the colors of the silver salts (AgCl, white; AgBr, pale yellow; and AgI, yellow) are indicative, confirmatory tests are desirable. Bromide and iodide ions can be confirmed by displacement reactions with free chlorine. The color of the liberated halogen is intensified with the use of a solvent such as 1,1,1-trichloroethane or chloroform.

$$2\,Br^- + Cl_2 \rightarrow Br_2 + 2\,Cl^-$$

$$2\,I^- + Cl_2 \rightarrow I_2 + 2\,Cl^-$$

Nitrate and Nitrite

If a nitrate is treated with concentrated H_2SO_4 followed by careful addition of a solution of iron(II) sulfate, a brown ring, known as nitrosyl ferrous sulfate ($FeSO_4 \cdot NO$), is formed at the junction of the sulfuric acid and iron(II) sulfate solutions. Sulfuric acid liberates nitric acid from the nitrate present. The HNO_3 is reduced by iron(II) sulfate—then the unstable complex $FeSO_4 \cdot NO$ is formed.

$$2\,HNO_3 + 6\,FeSO_4 + 3\,H_2SO_4 \rightarrow 3\,Fe_2(SO_4)_3 + 2\,NO + 4\,H_2O$$

$$FeSO_4 + NO \rightarrow \underset{\text{nitrosyl ferrous sulfate}}{FeSO_4 \cdot NO}$$

The test for a nitrite is similar to the test for nitrate except that acetic acid is used instead of sulfuric acid. The test for a nitrate will also yield a brown ring test with a nitrite; consequently, if a sample may contain both nitrite and nitrate, further tests are necessary to confirm a nitrate in the presence of a nitrite.

Acetate

The addition of concentrated H_2SO_4 and ethyl alcohol to an acetate, followed by warming, produces ethyl acetate, a volatile ester that is identified by its sweet and fruity odor. Sulfuric acid reacts with the acetate to give acetic acid, which then reacts with the alcohol as follows:

$$HC_2H_3O_2 + C_2H_5OH \rightarrow \underset{\text{ethyl acetate}}{C_2H_3O_2C_2H_5} + H_2O$$

An alternate test consists of testing an unknown solution with lanthanum nitrate and potassium iodide—iodine solutions. If an acetate is present, a basic lanthanum acetate is formed that will adsorb iodine to give a dark blue color. Furthermore, a dilute, slightly yellow solution of $FeCl_3$ will turn reddish when gently heated with an acetate.

Preliminary Experiment on Specific Tests

Carry out specific tests for all the anions listed above according to Procedure 3. Use test solutions of the anions or the solid salts, whichever are called for in the procedure. If there is any question about a test working out satisfactorily, consult the instructor. If solutions are not already made up, dissolve a small amount (volume of a small pea) of the sodium salt in 2 mL H_2O. Make notes on each specific test and write equations for the reactions.

Procedure 3 Specific Tests for Anions*

1. Sulfate: To 5 drops of solution, add 2 drops 6 *M* HCl and 2 drops $BaCl_2$ solution. A white precipitate ($BaSO_4$) proves SO_4^{2-}.

2. Arsenate: Arsenic is also determinable in cation analysis. To 5 drops of the prepared solution add 6 *M* NH_3 until basic; then add 3 or 4 drops of magnesia mixture. If a white crystalline precipitate forms, it may be either $MgNH_4PO_4$ or $MgNH_4AsO_4$.

Centrifuge the precipitate. Wash precipitate twice with 4 drops H_2O; then add 1 drop acetic acid and 1 drop silver nitrate solution to the precipitate. Formation of a yellow precipitate (Ag_3PO_4) indicates PO_4^{3-}; formation of a chocolate-colored precipitate proves AsO_4^{3-}.

3. Phosphate: Arsenates interfere with the following phosphate test. Unknown mixtures of phosphate and arsenic compounds will not be given. After proving arsenate absent (step 2 above), proceed.

To 5 drops of prepared solution add 5 drops dilute HNO_3 and 5 drops ammonium molybdate solution. Heat to boiling for several seconds, and allow to stand for a few minutes. Formation of a finely divided canary-yellow precipitate $[(NH_4)_3PO_4 \cdot 12\ MoO_3]$ proves PO_4^{3-}.

4. Borate: Make this test on the original sample. To about 0.1 g of the solid add a few drops concentrated H_2SO_4 in an evaporating dish. Mix thoroughly; add 2 or 3 mL methyl alcohol (or ethyl alcohol), heat, and then set fire to the alcohol. The immediate appearance of a green flame proves BO_3^{3-}.

5. Chromate: Chromium will also be found in cation analysis. To 5 drops of prepared solution add dilute NH_3 until basic, then dilute acetic acid until acid. Add 2 drops $BaCl_2$ solution. Formation of a yellow precipitate ($BaCrO_4$) proves CrO_4^{2-}.

To confirm CrO_4^{2-}, centrifuge and then wash the precipitate twice with 6 drops cold H_2O. To the precipitate add 2 drops dilute HNO_3 and 10 drops of H_2O; then add 4 or 5 drops ether and 2 drops of 3% H_2O_2 solution. Shake the tube well. A blue color appearing in the ether layer confirms CrO_4^{2-}.

6. Sulfite: Make this test on the original sample. To a small amount of solid in a small test tube, add 5 drops 6 *M* HCl, and heat gently. Effervescence and an odor of SO_2 indicates sulfite. (Do not mistake HCl odor for that of SO_2.)

To further test, add dilute HCl to 5 drops of solution or to a small amount of solid until distinctly acid, dilute with several drops H_2O, then add 2 drops $BaCl_2$. If a precipitate forms, centrifuge and discard precipitate ($BaSO_4$).

(continued)

*If these tests are being made on sodium salts, it is only necessary to dissolve the salt in water for preparing a solution. If cations other than Na^+, K^+, and NH_4^+ are present, it will be necessary to treat the sample according to Procedure 15 for preparing a solution for anion testing.

Add 4 drops 3% H_2O_2 to the filtrate. Formation of a white precipitate ($BaSO_4$) proves SO_3^{2-}.

7. Carbonate: Make this test on the original sample.

(a) *If sulfite is absent:* To a small amount of solid (volume of a small pea) in test tube add a few drops 6 *M* HCl. Test for evolution of CO_2 by using a small test tube generator. Fit a small test tube with a cork carrying a delivery tube that can be dipped into a $Ba(OH)_2$ solution in another small test tube. Add a few drops of concentrated HCl to about ¼ inch of solid in the generator, quickly replace the stopper, and allow the evolved gas to bubble into the $Ba(OH)_2$ solution. A white precipitate soon forms if carbonate is present. Any sample that effervesces with HCl to produce a colorless, odorless gas must contain a carbonate.

(b) *If sulfite is present:* To a small amount of solid in a test tube add an equal volume of Na_2O_2; then add a few drops of H_2O. Shake thoroughly. Add dilute HCl until acid, and proceed as in step 7 in testing for carbonate.

8. Sulfide: Make this test on the original sample. To a small amount of solid in a test tube add 5 drops 6 *M* HCl. Hold a strip of filter paper that has been moistened with lead nitrate solution over the mouth of the tube. If the paper is blackened (PbS), S^{2-} is present.

If the sample fails to dissolve by this treatment, add a few granules of zinc to the tube and warm. Test again with lead nitrate paper.

9. Chloride, Bromide, and Iodide: To 5 drops of prepared solution add 10 drops dilute HNO_3. Heat until there is no evidence of effervescence. (If a sulfite is present, it must be destroyed before proceeding.)

Add 1 drop silver nitrate solution. A white precipitate soluble in NH_3 indicates Cl^-; a pale yellow precipitate (AgBr) sparingly soluble in NH_3 indicates Br^-; a yellow precipitate (AgI) insoluble in NH_3 indicates I^-.

To detect bromide or iodide, add 6 drops chlorine water to 5 drops of prepared solution, followed by 10 drops $CHCl_3$ (see note 2). Shake the tube well. The $CHCl_3$ layer will be colored brown by bromine or violet by iodine. (To test for the presence of Cl^-, Br^-, and I^- in the presence of one another, see note 2.)

10. Nitrate: Make this test on the original sample. If unknown is a solid the test for a nitrate as on page 27 is conclusive, and further tests need not be made.

(a) *If bromide or iodide is absent:* Dissolve a pinch of the solid in a few drops of H_2O, or, in the event the sample is insoluble in H_2O, add 3 drops 6 *M* HCl. If there is any residue after acid treatment, centrifuge and discard. Add 10 drops concentrated H_2SO_4 to solution in test tube, then cool tube under tap. Now incline the tube at an angle of about 30° with the horizontal and add several drops

(continued)

Procedure 3 *(continued)*

of freshly prepared $FeSO_4$ solution *very slowly*, so that it forms a layer that floats on the H_2SO_4 solution. (The $FeSO_4$ solution may be prepared by dissolving 1 g of solid in 1 or 2 mL H_2O.) Bring the tube to an upright position and allow it to stand for about 1 minute. The formation of a brown ring ($FeSO_4 \cdot NO$) at the junction of the solutions proves NO_3^-.

(b) *If bromides or iodides are present:* Prepare a solution for the nitrate test as in **(a)** above; then add silver sulfate solution until precipitate is complete. Centrifuge and discard precipitate (AgBr, AgI). Test the filtrate as in **(a)** above with $FeSO_4$ and H_2SO_4.

11. Nitrite: Make this test on the original sample. Treat a small amount of the solid with a few drops H_2O. Reject any residue. Test the solution as in step 10, but use acetic acid instead of H_2SO_4. A deep brown color proves NO_2^-.

12. Acetate: Make this test on the original sample. Treat a small amount of solid with 6 drops concentrated H_2SO_4, followed by addition of 6 drops ethyl alcohol. Warm tube slowly to near boiling. A sweet or fruity odor (ethyl acetate) proves $C_2H_3O_2^-$.

Alternate test 1. **(a)** If the sulfate group is absent, dilute 2 or 3 drops of unknown solution to about 1 mL or, if sample is a solid, dissolve a small amount in H_2O. Add 4 drops $La(NO_3)_3$ solution (5%). If precipitate forms, add 1 or 2 drops dilute HNO_3 to redissolve. Add 1 drop $KI–I_2$ solution, make just alkaline with 6 *M* NH_3, and warm in water bath. A deep blue–black precipitate indicates acetate ion.

(b) If members of sulfate group are present, they must be removed. Dilute 2 or 3 drops of unknown solution to about 1 mL. Add 1 or 2 drops $BaCl_2$ solution and 6 *M* NH_3 until alkaline. Filter off any precipitate and reject it. Treat solution according to **(a)**.

Alternate test 2. Use 1 mL of dilute yellow $FeCl_3$ solution and add a small crystal of unknown solid or 4 drops of solution to it. Heat gently. A reddish color to the solution indicates acetate (3).

Notes

1. Some instructors may want to add tartrate as an example of an organic compound. In general, organic compounds char when heated with concentrated H_2SO_4.

Tartrate: **(a)** To a small quantity of $Na_2C_4H_4O_6$ in a small test tube, add a few drops of concentrated H_2SO_4. Heat gently, and note charring and some evolution of gas (SO_2).

(b) Silver mirror test. Clean a 6-inch test tube by boiling NaOH solution in it (CAUTION: very hazardous solution) and then washing it thoroughly with H_2O. Add 2 mL $AgNO_3$ to the tube; then add dilute NH_3 *drop by drop* with shaking until the precipitate that forms just dissolves or almost dissolves. Avoid excess NH_3 (it is best to stop adding NH_3 when a slightly gray precipitate of AgOH is still apparent). Add about 3 mL of the unknown solution and warm gently. Tartrates, if present, will reduce the silver compound to metallic silver, which will coat the inside of the test tube as a mirror.

$$AgNO_3 + NH_3 + H_2O \rightarrow AgOH \downarrow + NH_4NO_3$$

$$AgOH \rightarrow Ag^+ + OH^-$$

$$Ag^+ + 2\ NH_3 \rightarrow Ag(NH_3)_2^+$$

$$Ag(NH_3)_2^+ + Na_2C_4H_4O_6 \rightarrow Ag + 2\ NH_3 + \text{oxidation product}$$

2. Detection of Cl^-, Br^-, I^- in the presence of one another: In the brief scope of this course unknown mixtures of halogen salts will not be given. The literature has many references on methods of separation of the halogens, and thus separate identification. Tests given in the past few pages are adequate for the detection of iodide or bromide in chloride. An example of a method of removing bromine and iodine from chloride for follow-up chloride detection is as follows: A water solution of the three halide salts may be boiled with a few crystals of potassium dichromate. Iodine is liberated; on adding a few drops of H_2SO_4, Br_2 is liberated.

$$6\ I^- + Cr_2O_7^{2-} + 7\ H_2O \rightarrow 2\ Cr(OH)_3 + 3\ I_2 + 8\ OH^-$$

$$Cr_2O_7^{2-} + 6\ Br^- + 14\ H^+ \rightarrow 2\ Cr^{3+} + 3\ Br_2 + 7\ H_2O$$

Heating or chloroform extraction may be used to remove the free Br_2 or I_2 from the solution, which would still contain chloride ion.

All chlorinated organic solvents are toxic to some degree and should be handled very carefully. Carbon tetrachloride, CCl_4, was formerly used in these tests, but it has been banned from most first-year laboratories. Chloroform, $CHCl_3$, is somewhat less toxic. Better yet is 1,1,1,-trichloroethane, if available.

3. The *solution* should be red here. If anions are present that form insoluble iron(III) salts, a reddish rust-colored precipitate will be obtained; for example, CO_3^{2-} would precipitate iron(III) carbonate.

Analysis of Anion Unknowns

After completion of (1) action of concentrated H_2SO_4, (2) group tests, (3) specific tests on the sodium salts of the various anions, and (4) approval of written work on these, the student should analyze a group of unknowns, each of which is a pure sample of one sodium salt.

First try the action of concentrated H_2SO_4, Procedure 1. If this gives a good idea as to the nature of the anion, proceed immediately to specific tests, Procedure 3. If no good clues are obtained in the H_2SO_4 treatment, analyze according to Procedures 2 and 3. Record observations on all tests made on the unknowns. Each record should lead to conclusions as to the anion present.

Analysis of Anion Mixtures

When analyzing an unknown containing two or more anions, in general, the procedure given above for single anion unknowns should be followed. The action of sulfuric acid on the mixture should indicate a great deal. Follow with solubility group tests (Procedure 2) and specific tests (Procedure 3).*

Be sure to confirm evidence that suggests certain anions. For example, treatment of an iodide with concentrated H_2SO_4 suggests sulfite and sulfide as well as iodide, since the latter reduces H_2SO_4 to SO_2 and H_2S. However, if S^{2-} and SO_3^{2-} are absent in the original sample, this will be indicated by confirmatory tests for these anions.

---

*It is true that certain anions interfere with the testing for others, but it is beyond the scope of this course to consider all such problems. Instructors should issue anion mixtures that will be compatible with the outlined procedures of analysis.

PART THREE

ANALYSIS FOR CATIONS

Cation Group Separations*

Ag^{+}, Pb^{2+}, Hg_2^{2+}
Group I

Hg^{2+}, Cu^{2+}, Bi^{3+}, Cd^{2+}, As^{3+}, AsO_4^{3-}, Sb^{3+}, SbO_4^{3-}, Sn^{2+}, Sn^{4+}
Group II

Fe^{2+}, Fe^{3+}, Al^{3+}, Cr^{3+}, Zn^{2+}, Co^{2+}, Ni^{2+}, Mn^{2+}
Group III

Ba^{2+}, Sr^{2+}, Ca^{2+}
Group IV

Mg^{2+}, Na^{+}, K^{+}, Li^{+}, NH_4^{+}
Group V

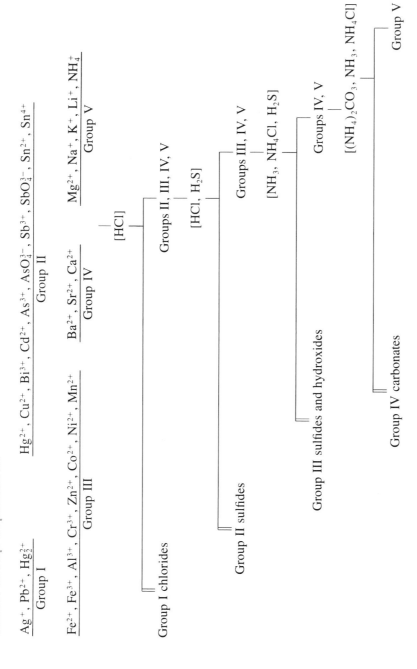

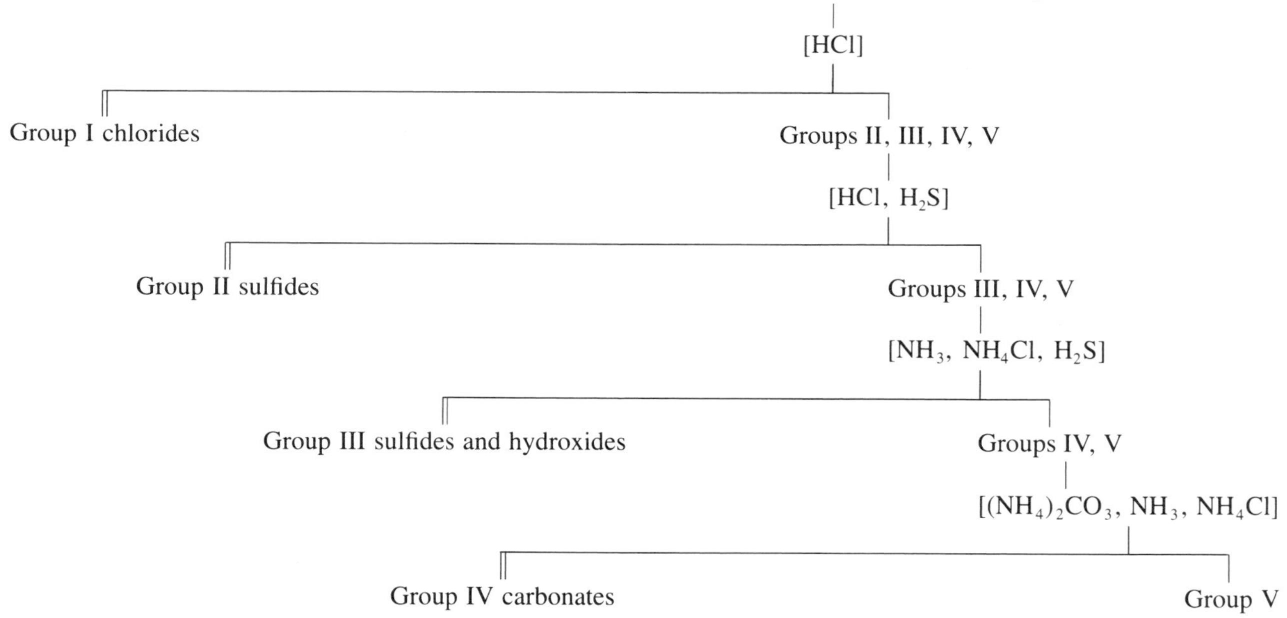

*In the above diagram of the cation group separations and in similar diagrams for other groups,

——— a horizontal line indicates a filtration or separation by centrifuge,

denotes a precipitate,

denotes a filtrate, and

[ ] denotes an added reagent.

INTRODUCTION TO CATION ANALYSIS

9

Among the problems confronting the analyst are (1) identification of metals, (2) analysis of alloys for metallic components, and (3) detection of metals as constituents of compounds. In the case of water solutions of compounds of metals, the problem is one of *identification of metallic ions* (cations), which is the subject of this section. Dissolving metals and alloys to obtain ions and the solution of compounds insoluble in water will be discussed later. In the brief course of qualitative analysis for which this manual is designed, only the more common or representative cations are included for analysis.

Method of Cation Analysis

Some few cations may be detected in the presence of several other cations, but the usual procedure for analyzing a mixture of several cations involves a *systematic separation of cations into groups, followed by separation of each group into its components*. By taking advantage of differences in chemical properties, groups of ions are separated from other ions in solution through *group reagents* that form precipitates with given groups. To illustrate, if hydrochloric acid is added to a solution containing all cations, only the chlorides of lead, silver, and mercury(I) precipitate, since all other chlorides are soluble. (See solubility rules, page 9.) Lead, silver, and mercury(I) ions then constitute a group of ions that can be precipitated from solution by addition of the group reagent HCl. After removal of this group of ions, another group can be precipitated by adding hydrogen sulfide to the acidified solution. By treating the resulting solution, other groups can be separated by addition of group reagents; that is, the cations are divided into five groups based on differences in chemical properties:

Group I	Ag^+, Pb^{2+}, Hg_2^{2+}
Group II	Pb^{2+}, Hg^{2+}, Cu^{2+}, Bi^{3+}, Cd^{2+}, As^{3+}, AsO_4^{3-}, Sb^{3+}, SbO_4^{3-}, Sn^{2+}, Sn^{4+} (Arsenic and antimony probably do not exist as simple ions when exhibiting oxidation numbers of 5.)
Group III	Fe^{2+}, Fe^{3+}, Al^{3+}, Cr^{3+}, Ni^{2+}, Co^{2+}, Mn^{2+}, Zn^{2+}
Group IV	Ba^{2+}, Sr^{2+}, Ca^{2+}
Group V	Mg^{2+}, Na^+, K^+, Li^+, NH_4^+

The separation of the five groups is shown diagrammatically on page 34.

After the ions have been separated into groups, it is necessary to separate the ions of each group; these separations are also dependent on differences in chemical properties of the constituent ions. For example, in Group I, lead chloride can be separated from silver and mercury(I) chlorides because lead chloride is soluble in hot water, whereas the chlorides of mercury and silver are relatively insoluble. The chlorides of mercury and silver then can be separated with ammonia, since silver chloride is soluble and mercury(I) chloride is insoluble in that reagent. After the ions have been separated from one another, *specific tests* are applied; that is, a reagent is added that gives a characteristic precipitate or color with the ion in question.

In general, the procedure for each of the cation unknowns will be: (1) make a group separation, (2) separate the ions in each group, and (3) after separation, make specific tests for the cations of each group. The chemical basis for each of the group separations and specific tests will be considered at the appropriate place in the qualitative procedure.

Study the diagram of analysis shown on page 34 carefully, for similar diagrams are to be developed as each sample is analyzed. **The principal purpose of such an outline is to record steps taken in the analysis. Each step taken in the analytical procedure is to be recorded in this outline at the time of doing the experimental work. Colors of precipitates and colors of solutions should be carefully noted on the diagram, since these colors are often indicative of the presence of certain ions in the sample.** Before any sample is reported, a careful check of the diagram should be made to see that the ions to be reported conform in properties to the notes made. **Such a schematic representation of the analysis is often required with every report of analysis.**

In general, the student will first analyze a known solution (that is, a solution known to contain certain specific ions) before proceeding to the analysis of an unknown sample. The purpose of the method is to familiarize the student thoroughly with the techniques involved in the group separations and final tests before applying them to an unknown. A series of preliminary tests for the individual ions precedes each of the group analyses. These may be performed at the option of the instructor.

PRELIMINARY EXPERIMENT ON GROUP SEPARATIONS

10

This experiment illustrates the use of group reagents in separating the cations into five major divisions. One cation from each group has been chosen, and a simple procedure is given for each separation. As the procedure is carried out, the student should develop a diagram for the steps similar to the one on page 34, and write ionic equations for the precipitation of the cations.

Preliminary Experiment

Place 2 drops of each of the following test solutions in a small test tube: Ag^+, Cu^{2+}, Ni^{2+}, Ba^{2+}, and K^+. Add 10 drops of H_2O and 2 drops of 1 *M* HCl. Stir, heat, and then centrifuge. A precipitate indicates an element of Group I. Test for complete precipitation. Decant or pipet the solution to another tube (1).

Adjust the hydrogen ion concentration for Group II precipitation as follows: Add dilute NH_3 drop by drop until the solution shows violet on methyl violet paper (2). (Use a wire or glass rod to wet a small spot on the paper.) Acidify the solution with 6 *M* HCl until just green to methyl violet paper. Add 2 drops of NH_4Cl solution.

Saturate the solution with H_2S by adding 6 drops of thioacetamide solution. Then heat the tube in a boiling water bath for at least 5 minutes. The formation of a precipitate indicates a cation of Group II. Centrifuge. If the solution is not clear, bring it to a boil and centrifuge again. Test for complete precipitation.

Transfer the solution to a crucible and heat to boiling to remove H_2S (3). Add concentrated NH_3 until the solution is basic to litmus. Transfer the mixture to a test tube and saturate with H_2S by adding 6 drops thioacetamide and heating in a boiling water bath for 5 minutes. Precipitation indicates a cation of Group III. Centrifuge.

Remove the supernatant liquid; then evaporate the solution to a volume of about 1 mL. Centrifuge again if not clear.

(continued)

Preliminary Experiment *(continued)*

Add 2 drops NH_4Cl solution, make basic with NH_3, and add 2 drops $(NH_4)_2CO_3$ solution. Heat to boiling. Precipitation indicates a cation of Group IV.

Centrifuge. The presence of a cation of Group V can be proved only by making specific tests on the filtrate.

Notes

1. This is a simplified procedure adapted to the five ions present. In a general unknown, the procedure under each group should be followed.

2. The proper acidity can also be obtained without the use of methyl violet paper in the following manner: Carefully neutralize the solution by adding 6 *M* NH_3 dropwise until the solution tests just basic with litmus paper. Then add *exactly* 2 drops of 6 *M* HCl.

3. If the test tube is not over half full, the boiling can be done in the tube if it is heated very cautiously. If boiled in a crucible, transfer back to a test tube before proceeding.

GROUP I: THE SILVER GROUP

11

Group I: The Silver Group

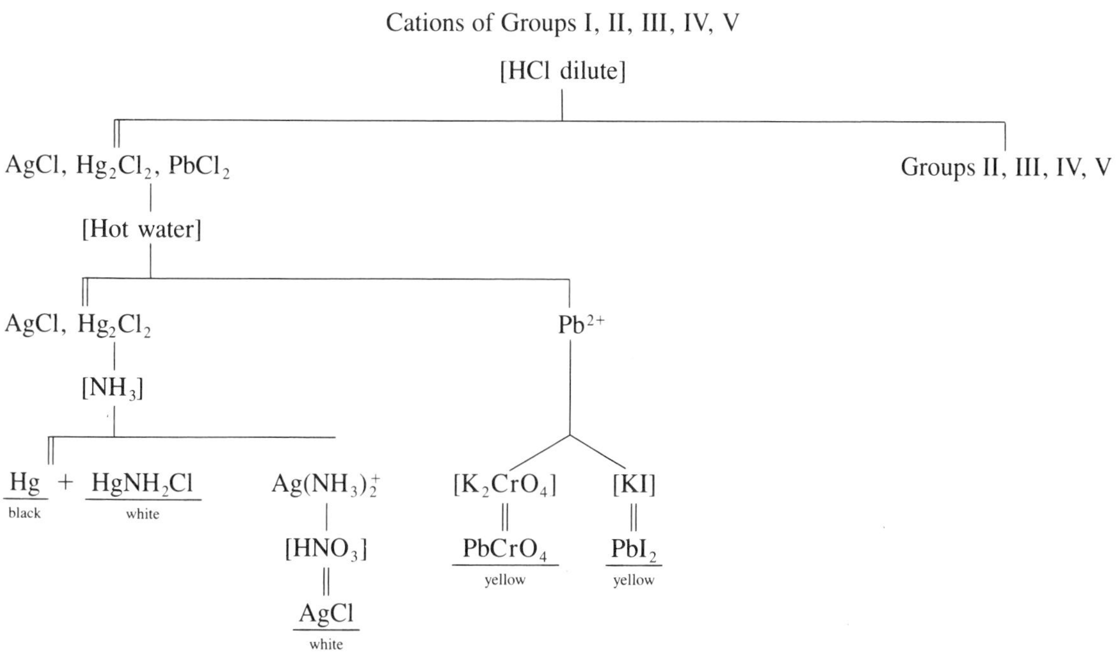

Chemistry of Separations and Tests

1. The fact that chlorides of Ag^+, Pb^{2+}, and Hg_2^{2+} are insoluble is the basis for the separation of Group I from Groups II, III, IV, and V. Since only Group I chlorides are insoluble, addition of chloride ion (HCl) precipitates AgCl, $PbCl_2$, and Hg_2Cl_2 from a solution containing all of the cations. Refer to the diagram of the procedure on this page.

$$Ag^+ + Cl^- \rightarrow \underline{AgCl}$$

$$Pb^{2+} + 2\ Cl^- \rightarrow \underline{PbCl_2}$$

$$Hg_2^{2+} + 2\ Cl^- \rightarrow \underline{Hg_2Cl_2}$$

2. $PbCl_2$ is soluble in hot water; hence it can be separated from AgCl and Hg_2Cl_2, which remain insoluble. Lead is identified by precipitation of yellow lead chromate

$$Pb^{2+} + CrO_4^{2-} \rightarrow \underline{PbCrO_4}$$

or precipitation of orange-yellow lead iodide.

3. AgCl can then be separated from Hg_2Cl_2 because of its solubility in NH_3 to form the complex soluble ion $Ag(NH_3)_2^+$, called diammine silver ion.

$$\underline{AgCl} + 2\ NH_3 \rightarrow Ag(NH_3)_2^+ + Cl^-$$

4. Hg_2Cl_2 is converted to insoluble Hg and $HgNH_2Cl$ (amidomercury(II) chloride) by its action with NH_3.

$$Hg_2Cl_2 + 2\ NH_3 \rightarrow \underset{\text{black}}{\underline{Hg}} + \underset{\text{white}}{\underline{HgNH_2Cl}} + NH_4^+ + Cl^-$$

In this equation one mercury(I) atom is oxidized to oxidation number 2, while the other is reduced to atomic mercury of zero oxidation state. The black precipitate produced serves as a test for the presence of mercury(I).

5. Silver is identified by reprecipitation of AgCl from the ammoniacal solution with H^+ (using HNO_3).

$$Ag(NH_3)_2^+ + Cl^- + 2\ H^+ \rightarrow \underline{AgCl} + 2\ NH_4^+$$

Preliminary Tests for Group I

Record observations and write equations for all chemical changes.

1. Solubility of Group I chlorides in water. Place 3 drops of Pb^{2+} test solution in a test tube and add 1 drop of 6 *M* HCl. Cool under the tap and let stand for a few minutes. Heat to near boiling and observe any changes.
Repeat the procedure, using 1 drop of Hg_2^{2+} test solution and again with 1 drop of Ag^+ test solution. Centrifuge and save precipitates for subsequent tests.

2. Precipitation of $PbCrO_4$. To 1 drop of Pb^{2+} add 1 drop of K_2CrO_4. What change takes place?

3. Action of ammonia on Hg_2Cl_2 and AgCl. To the tubes containing the Hg_2Cl_2 and AgCl precipitates from (1), add 15 *M* NH_3 dropwise and observe changes. To the tube that contained AgCl add 6 *M* HNO_3 until acid.*

Analytical Procedure for Group I

To become thoroughly familiar with the separations and tests for ions of this group, the student should make up a sample containing all of the ions of the group (called a known solution) and analyze according to Procedure 4 below. For the known sample use

*Throughout qualitative analysis, it is important to check acidity or basicity often. In this case, a few drops of HNO_3 might seem to be enough, but since 15 *M* NH_3 was used, an unexpectedly larger amount of HNO_3 is needed to render the solution slightly acid. Errors are made at times by not stirring a column of liquid in a small test tube: the upper part may be acid and the lower part basic, and vice versa.

2 drops each of Ag^+ and Hg_2^{2+} and 10 drops of Pb^{2+} test solutions. Save the specific or final tests to show to the instructor.

Procedure 4 Analysis of Silver Group
Solution: (1) Place 0.5 mL of the solution to be analyzed in a 3- or 4-inch test tube. Add 1 drop 6 *M* HCl (2). Heat the tube by inserting into a bath of hot water, in order to coagulate the precipitate. Cool under the tap and allow to stand 3 minutes (3). Centrifuge. Test for complete precipitation by adding a drop of 6 *M* HCl to the supernatant liquid in the tube. Pipet or decant the solution into a test tube. Wash the precipitate with several drops 1 *M* HCl (4) and add washings to the solution already drawn off.
Solution: Groups II, III, IV, V. Analyze according to Procedure 5.
Precipitate: AgCl, Hg_2Cl_2, $PbCl_2$. Add 8 drops H_2O, heat the tube to near boiling in a water bath, and shake to digest the precipitate (5). Centrifuge. Before cooling, decant the solution. Wash the precipitate with 4 drops more of hot water. Combine the washings (6).
Solution: Pb^{2+}. To 2 drops of the solution in a test tube add 1 drop of K_2CrO_4. Formation of yellow precipitate indicates Pb^{2+} (7).
Precipitate: Hg_2Cl_2, AgCl. Add 4 drops of concentrated NH_3. Formation of a black precipitate proves mercury. Shake the tube. Centrifuge. Decant the solution. Repeat the extraction with 4 drops more of NH_3 and combine the two extractions.
Solution: $Ag(NH_3)_2^+$. To the solution in a test tube add dilute HNO_3 drop by drop until the solution is definitely acid to litmus. White precipitate confirms *silver*.
Precipitate: $HgNH_2Cl$, Hg.

Notes

1. A known for any group should contain about 1.0 mg of each ion. It is prepared by mixing 2 drops of test solution of each of the ions to be tested, except lead. Ten drops of Pb^{2+} test solution should be used because of the relatively higher solubility of $PbCl_2$.

When an unknown solution is received from the instructor, use only 0.5 mL for analysis. Reserve the rest for repeating doubtful tests.

2. A large excess of HCl will partially dissolve AgCl and Hg_2Cl_2.

3. $PbCl_2$ is slow to precipitate. If only Group I is to be tested, run a test on the filtrate for Pb^{2+}.

4. 1 *M* HCl will dissolve SbOCl and BiOCl, which may have precipitated if present.

5. The use of a small beaker of water for a water bath is the safest manner of heating to boiling; however, if the volume is less than 1 mL, heating can be done directly over a micro flame.

6. If a large amount of $PbCl_2$ is present here, as might be obtained from many alloys, several washings may be necessary to completely remove $PbCl_2$. If some $PbCl_2$ remains here, it will not interfere in the test for Hg_2^{2+} since, if the latter is present, a black precipitate will be obtained on addition of NH_3. White $PbCl_2$ remains unchanged.

7. An alternate test for lead is the formation of an orange-yellow precipitate with dilute KI solution.

Analysis of Group I Unknown

After completion of the known sample, write equations for all reactions involved in the procedure. Group together all reactions of each cation of Group I; these equations should be in ionic form. After you have demonstrated familiarity with the group procedures, the instructor will issue an unknown, which may contain any or all three of the ions. This unknown is to be analyzed in the same manner as was the known solution, according to Procedure 4. The only written report required for an unknown is the diagram of analysis such as the one on page 39 that is *to be developed while making the analysis*. In this diagram, show all operations performed and all colors of solutions and precipitates.

GROUP II: THE COPPER–ARSENIC GROUP

12

Chemistry of Separations and Tests

Precipitation and separation of copper and arsenic subgroups

1. When H_2S is added to a 0.3 *M* HCl solution containing cations of Groups II, III, IV, and V, only the sulfides of Group II are precipitated, because the sulfides of cations of Groups III, IV, and V are soluble in 0.3 *M* HCl. The sulfides precipitate in approximately the following order: As_2S_5, As_2S_3, HgS, CuS, Sb_2S_3, Sb_2S_5, SnS_2, Bi_2S_3, PbS, CdS, SnS. An example of one of the precipitation reactions would be

$$Cu^{2+} + H_2S \rightarrow \underline{CuS} + 2\ H^+$$

(Refer to Section 18.3 of the text for application of the solubility product principle to the precipitation of cations of Group II.)

2. The basis for the separation of Hg, As, Sb, Sn from Cu, Pb, Bi, Cd is that the sulfides of the arsenic subgroup are soluble in sodium polysulfide solution (Na_2S_2), whereas the sulfides of the copper group are insoluble. (Refer to diagram on page 44.) Treatment with Na_2S_2 oxidizes any lower valence As, Sb, Sn to the high oxidation states, for example,

$$As_2S_3 + 2\ S_2^{2-} \rightarrow As_2S_5 + 2\ S^{2-}$$

$$SnS + S_2^{2-} \rightarrow SnS_3^{2-}$$

The polysulfide solution contains sulfide ions (S^{2-}) as well as disulfide ions (S_2^{2-}). The former react with the high-valence sulfides to form thio salts, for example,

$$As_2S_5 + 3\ S^{2-} \rightarrow 2\ AsS_4^{3-}$$

$$SnS_2 + S^{2-} \rightarrow SnS_3^{2-}$$

AsS_4^{3-} is the thioarsenate ion, and SnS_3^{2-} the thiostannate ion.

Analysis of copper subgroup

3. The sulfides of the copper group (CuS, PbS, Bi_2S_3, CdS) are readily soluble in hot dilute HNO_3, for example,

$$3\ CuS + 8\ H^+ + 2\ NO_3^- \rightarrow 3\ Cu^{2+} + 2\ NO + 2\ H_2O + 3\ S$$

4. When a solution of the cations (Cu^{2+}, Pb^{2+}, Bi^{3+}, Cd^{2+}) has been obtained, lead

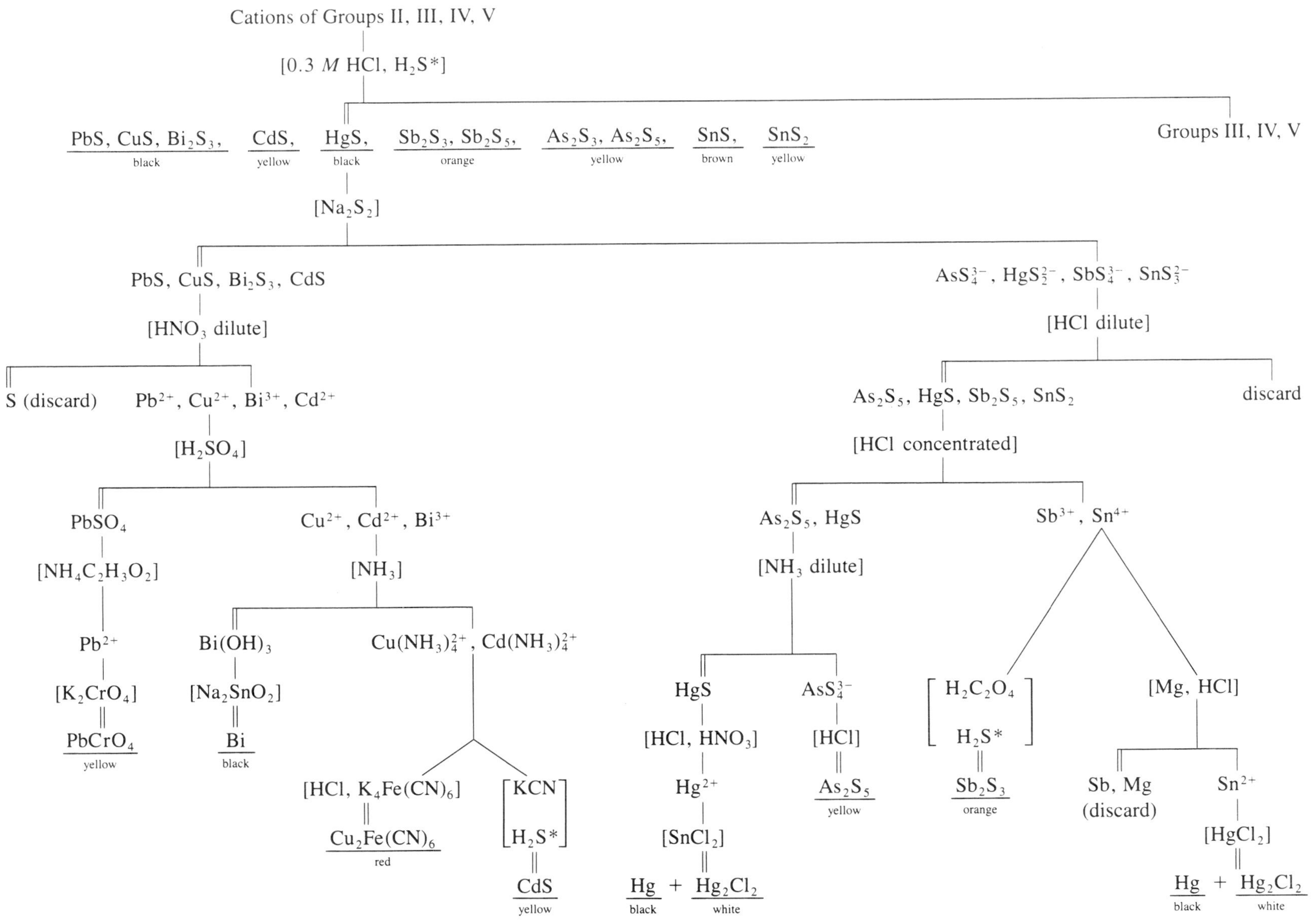

*Or thioacetamide.

may be precipitated as $PbSO_4$ by the addition of sulfate ion (H_2SO_4); sulfates of the three other cations are soluble, and therefore are not precipitated.

$$Pb^{2+} + SO_4^{2-} \rightarrow \underline{PbSO_4}$$

In the identification of lead, $PbSO_4$ is dissolved in $NH_4C_2H_3O_2$ solution, and Pb^{2+} is precipitated as yellow $PbCrO_4$ on addition of K_2CrO_4. $PbSO_4$ dissolves in ammonium acetate because $Pb(C_2H_3O_2)_2$ is almost completely nonionized.

$$PbSO_4 \rightleftharpoons Pb^{2+} + SO_4^{2-}$$
$$2\ C_2H_3O_2^-$$
$$\downarrow$$
$$Pb(C_2H_3O_2)_2$$

5. The basis for the separation of Bi^{3+} from Cd^{2+} and Cu^{2+} is the formation of insoluble $Bi(OH)_3$ on the addition of NH_3 to a solution of the three ions.

$$NH_3 + H_2O \rightleftharpoons NH_4^+ + OH^-$$

$$Bi^{3+} + 3\ OH^- \rightarrow \underline{Bi(OH)_3}$$

While $Cu(OH)_2$ and $Cd(OH)_2$ are insoluble in water, both are soluble in excess NH_3 solution because of the formation of complex ions.

$$Cu^{2+} + 4\ NH_3 \rightarrow \underset{\text{blue}}{Cu(NH_3)_4^{2+}}$$

$$Cd^{2+} + 4\ NH_3 \rightarrow \underset{\text{colorless}}{Cd(NH_3)_4^{2+}}$$

Bismuth forms no complex ion with NH_3; hence the hydroxide remains insoluble in excess NH_3. Bismuth can be confirmed by the addition of sodium stannite solution to the precipitate of $Bi(OH)_3$. $Bi(OH)_3$ is instantly reduced to a finely divided black precipitate of Bi.

$$2\ Bi(OH)_3 + 3\ SnO_2^{2-} \rightarrow 2\ \underline{Bi} + 3\ SnO_3^{2-} + 3\ H_2O$$

6. The intense blue color of $Cu(NH_3)_4^{2+}$ is evidence of the presence of Cu^{2+}. Copper can be confirmed by precipitation of red-brown copper(II) ferrocyanide from an acid solution with ferrocyanide ion.

$$Cu(NH)_3)_4^{2+} + 4\ H^+ \rightarrow Cu^{2+} + 4\ NH_4^+$$

$$2\ Cu^{2+} + Fe(CN)_6^{4-} \rightarrow \underset{\text{red-brown}}{\underline{Cu_2Fe(CN)_6}}$$

7. Cadmium is identified by precipitation of CdS from a cyanide solution with H_2S. Cyanide ion forms a complex ion with Cd^{2+}, which dissociates to give sufficient Cd^{2+} for formation of CdS with H_2S.

$$Cd(NH_3)_4^{2+} + 4\ CN^- \rightarrow Cd(CN)_4^{2-} + 4\ NH_3$$

$$Cd(CN)_4^{2-} \rightleftharpoons Cd^{2+} + 4\ CN^-$$

$$Cd^{2+} + H_2S \rightarrow \underset{\text{yellow-orange}}{\underline{CdS}} + 2\ H^+$$

Copper ion is reduced in cyanide solution and remains as part of a copper(I) cyanide complex ion that ionizes so slightly that it is not precipitated by H_2S and does not interfere with yellow CdS precipitation.

$$2\ Cu(CN)_4^{2-} \rightarrow 2\ Cu(CN)_2^- + (CN)_2 + 2\ CN^-$$

(CAUTION: Never add cyanide to an acid solution! Lethal fumes of HCN can be generated from cyanide compounds in acid solution.)

Analysis of the arsenic subgroup

When acid is added to the solution of complex thio salts of the arsenic group, the members of the group are reprecipitated as sulfides.

$$2\ AsS_4^{3-} + 6\ H^+ \rightarrow \underline{As_2S_5} + 3\ H_2S \uparrow$$

$$2\ SbS_4^{3-} + 6\ H^+ \rightarrow \underline{Sb_2S_5} + 3\ H_2S \uparrow$$

$$SnS_3^{2-} + 2\ H^+ \rightarrow \underline{SnS_2} + H_2S \uparrow$$

$$HgS_2^{2-} + H^+ \rightarrow \underline{HgS} + H_2S \uparrow$$

Note that all the members of this group are reprecipitated in their higher oxidation states.

8. The basis for the separation of Sb and Sn from Hg and As is the solubility of the sulfides of the former two elements in concentrated HCl and the insolubility of HgS and As_2S_5 in that reagent. Treatment of the sulfides of the group with concentrated HCl gives the reactions

$$SnS_2 + 4\ H^+ \rightarrow Sn^{4+} + 2\ H_2S$$

$$Sb_2S_5 + 6\ H^+ \rightarrow 2\ Sb^{3+} + 3\ H_2S + 2\ S$$

9. The resulting solution of $SnCl_4$ and $SbCl_3$ can be tested for Sb^{3+} without removing Sn^{4+} by adding oxalic acid and H_2S. Antimony is precipitated as the orange sulfide.

$$2\ Sb^{3+} + 3\ H_2S \rightarrow \underset{\text{orange-red}}{\underline{Sb_2S_3}} + 6\ H^+$$

Sn^{4+} forms a complex ion with oxalic acid and is not precipitated.

10. In testing for tin it is necessary to reduce tin(IV) ion to tin(II) ion. This may be accomplished by the addition of iron, aluminum, or magnesium.

$$Sn^{4+} + Fe \rightarrow Sn^{2+} + Fe^{2+}$$

$$Sn^{4+} + Mg \rightarrow Sn^{2+} + Mg^{2+}$$

Antimony in the solution is simultaneously reduced to the free element, which appears as a flaky black residue.

$$2\ Sb^{3+} + 3\ Fe \rightarrow 3\ Fe^{2+} + 2\ \underline{Sb}$$

$$2\ Sb^{3+} + 3\ Mg \rightarrow 2\ \underline{Sb} + 3\ Mg^{2+}$$

After obtaining a solution of tin in the 2+ state, addition of $HgCl_2$ solution produces a grey precipitate composed of Hg (black) and Hg_2Cl_2 (white).

$$Hg^{2+} + Sn^{2+} \rightarrow \underline{Hg} + Sn^{4+}\ (\text{excess } SnCl_2)$$

$$2\ Cl^- + 2\ Hg^{2+} + Sn^{2+} \rightarrow \underline{Hg_2Cl_2} + Sn^{4+}\ (\text{excess } HgCl_2)$$

The relative amounts of Hg and Hg_2Cl_2 (and therefore the color of the precipitate) depend on the relative amounts of $SnCl_2$ and $HgCl_2$ present in the test.

11. The basis of the separation of As_2S_5 and HgS is the insolubility of HgS in NH_3. As_2S_5 dissolves according to the equation

$$As_2S_5 + 6\ NH_3 + 3\ H_2O \rightarrow (NH_4)_3AsO_3S + (NH_4)_3AsS_4$$

12. The test for arsenic consists of the reprecipitation of As_2S_5 from the ammoniacal solution with hydrochloric acid.

$$(NH_4)_3AsO_3S + (NH_4)_3AsS_4 + 6\ HCl \rightarrow \underset{\text{yellow}}{\underline{As_2S_5}} + 6\ NH_4Cl + 3\ H_2O$$

13. The test for mercury is made by dissolving HgS in aqua regia.

$$HgS + 4\,H^+ + NO_3^- + Cl^- \rightarrow Hg^{2+} + NOCl + 2\,H_2O + S$$

Addition of $SnCl_2$ to the solution results in the precipitation of Hg and Hg_2Cl_2. Reactions are the same as given in the test for tin.

Preliminary Tests for Group II

Record your observations and write the corresponding chemical equations.

1. Colors of sulfides. Put 2 drops of test solution of each ion of Group II (Hg^{2+}, Pb^{2+}, Bi^{3+}, Cd^{2+}, Cu^{2+}, As^{3+}, Sb^{3+}, Sn^{2+}) into as many separate test tubes. Dilute each to 0.5 mL, add 6 drops of thioacetamide solution, and heat in a boiling water bath for 5 minutes. After noting the colors of the precipitates, centrifuge and discard the supernatant liquids. Save the precipitate for the next part of this experiment. (Arsenic precipitates best in a hot HCl solution.)

2. Solubility of sulfides. To each of the above precipitates add 2 drops of sodium polysulfide, Na_2S_2, and 2 drops H_2O. Stir and warm. Record which dissolves rapidly, which slowly, and which not at all.

3. Methyl violet color changes. By dilution of 1 *M* HCl in a 10-mL graduate, prepare 0.5, 0.3, and 0.1 *M* solutions. Place some of each in as many test tubes, and place H_2O in a fourth. With a stirring rod add a drop of each to methyl violet papers and record the colors produced.

4. Hydrolysis of bismuth and antimony salts. To 1 drop of Bi^{3+} test solution add water drop by drop until precipitation occurs. Will the precipitate dissolve in 1 *M* HCl? Repeat with Sb^{3+}.

5. Precipitation and dissolving of lead sulfate. To 2 drops Pb^{2+} test solution add 1 drop dilute H_2SO_4 and 5 drops of H_2O. Centrifuge and discard liquid. Dissolve precipitate in a few drops of ammonium acetate solution. Warm if necessary. Add a drop of potassium chromate.

6. Reduction of bismuth hydroxide by stannite ion (SnO_2^{2-}). To 1 drop of Bi^{3+} test solution in a test tube, add 1 drop of dilute NH_3. In another test tube place 2 drops of $SnCl_2$ and add NaOH to it drop by drop until the precipitate first formed dissolves. Now add 1 drop of this solution to the mixture in the other test tube.

Analytical Procedures for Group II

The following procedures are to be used for analyzing known and unknown solutions. Prepare a known sample by mixing 2 drops of each test solution of ions of this group. (Use 4 drops each of Cd^{2+}* and Pb^{2+} test solutions; use Sn^{4+} rather than Sn^{2+}, since the latter would reduce Hg^{2+}.)

Analyze according to Procedures 5–7 below.

*Since Cd^{2+} is one of the more difficult to precipitate, it may be omitted at the option of the instructor.

Procedure 5 Precipitation and Separation of Copper and Arsenic Subgroups

Solution: (From Procedure 4, or sample known to contain Group II.) Adjust the volume of solution to about 1 mL by evaporation or addition of H_2O (1). Neutralize the solution by adding concentrated NH_3 dropwise until just basic to litmus. Now adjust the acidity of the solution to 0.3 *M* by adding 1 *M* HCl dropwise until a blue-green color is obtained with methyl violet test paper (2). (The acidity may be adjusted according to Note 3.)

Saturate the solution with H_2S.* Centrifuge and, if the solution is not clear, add 2 drops NH_4Cl solution and 2 drops H_2O (4), (5). Heat to boiling and centrifuge again.

Solution: Boil the solution to remove H_2S. If there is a residue, centrifuge and discard the precipitate. Treat the solution for Groups III, IV, and V according to Procedure 8.

Precipitate: (6) HgS (black), CuS (black), Bi_2S_3 (brown-black), PbS (black), CdS (yellow), As_2S_3–As_2S_5 (yellow), SnS (brown-black), SnS_2 (yellow), Sb_2S_3–Sb_2S_5 (orange). Wash the precipitate with 10 drops of H_2O and discard the washings.

To the precipitate add 3 drops Na_2S_2 and 7 drops H_2O. Warm and stir for 1–2 minutes. Centrifuge. Decant or draw off the solution. Repeat the extraction of the precipitate with Na_2S_2 as above and combine the two extraction solutions. Wash the precipitate once with 6 drops H_2O, combining the washing with the two extraction solutions.

Solution: Arsenic subgroup. Treat according to Procedure 7.

Precipitate: Copper subgroup. Treat the precipitate according to Procedure 6.

*Thioacetamide may be used as a source of H_2S in any or all of the procedures. Whenever saturation with H_2S is called for, simply add 6 to 8 drops of 8% thioacetamide solution, place the test tube in a beaker of boiling water for 5 minutes, and proceed according to directions.

Notes

1. If the solution from the silver group precipitate is concentrated here, some SbOCl or BiOCl may precipitate. Such a precipitate, however, should not be separated, since these substances are sufficiently soluble so that they will be converted to sulfides with H_2S.

2. If the solution is less than 0.1 *M* acid, some ZnS may precipitate in this group; if the acidity is too high, PbS and CdS may fail to precipitate. Methyl violet shows the following color changes:

Solution	Color
Neutral and alkaline	Violet
0.1 *M* H^+	Blue
0.3 *M* H^+	Blue-green

Solution	Color
0.5 *M* H^+	Yellow-green
1.0 *M* H^+	Yellow

Commercial *p*H paper may be used if available in this range. An acidity of 0.3 *M* corresponds to a *p*H of 0.5.

3. The proper acidity may be obtained here without the use of methyl violet paper if, after the solution has been carefully neutralized with NH_3, *exactly* 2 drops of 6 *M* HCl are added.

4. Dilution of the sample reduces the H^+ concentration, which facilitates precipitation of the more soluble sulfides (CdS, SnS, PbS).

5. Addition of ammonium ion helps coagulate colloidal sulfides. Formation of colloidal sulfides can at times be troublesome. The use of a few granules of solid NH_4Cl to furnish ions to help coagulate the colloid, and patient heating with stirring, usually renders the suspension susceptible to settling on centrifugation.

6. The color of the precipitate at this point may be indicative of ions present. A yellow precipitate would eliminate mercury, copper, bismuth, and lead, which form black sulfides.

Procedure 6 Analysis of Copper Subgroup

Precipitate: (From Procedure 5—CuS, PbS, Bi_2S_3, CdS, HgS.) Add 6 drops 6 *M* HNO_3 and 7–10 drops H_2O (1). Heat to boiling and keep the solution hot for 1–2 minutes. Centrifuge. If any appreciable amount of precipitate remains, again treat with HNO_3 and combine the two solutions.

Precipitate: S, HgS. Discard (2).

Solution: Pb^{2+}, Cu^{2+}, Bi^{3+}, Cd^{2+}. Transfer to a crucible or small casserole, add 3–4 drops concentrated H_2SO_4 (3). Evaporate cautiously until dense white fumes appear. Cool.

Add 10 drops H_2O (5). Stir thoroughly. Finally, transfer to a test tube and centrifuge. A white precipitate indicates lead. Decant the liquid. Wash precipitate with 5 drops H_2O and 1 drop dilute H_2SO_4. Add the washings to the liquid previously decanted.

Precipitate: $PbSO_4$ (white). Add 3 drops hot $NH_4C_2H_3O_2$ to the precipitate (4). Stir thoroughly. Add 1 drop K_2CrO_4. Formation of a yellow precipitate confirms *lead* (6).

Solution: Bi^{3+}, Cu^{2+}, Cd^{2+}. Add 15 *M* NH_3 dropwise until basic to litmus. A blue color indicates copper (7); a white flocculent precipitate indicates bismuth. Centrifuge.

(continued)

Procedure 6 *(continued)*

Precipitate: $Bi(OH)_3$ (white). Prepare Na_2SnO_2 by placing 2 drops $SnCl_2$ in test tube and adding NaOH until the precipitate formed dissolves (8); this may require 8–12 drops NaOH. Add this solution to the preceding precipitate. Formation of a black precipitate within a few seconds confirms *bismuth*.

Solution: $Cu(NH_3)_4^{2+}$ (blue), $Cd(NH_3)_4^{2+}$. Concentrate to about 1 mL.

(a) To 4 drops of the solution add 6 *M* HCl until acid; then 2 drops $K_4Fe(CN)_6$. Centrifuge. A red precipitate $[Cu_2Fe(CN)_6]$ confirms *copper*.

(b) To the remainder of solution, add KCN dropwise until blue color disappears, then 2 drops in excess. (Add 1 drop KCN if Cu was absent.) Saturate with H_2S (by using thioacetamide). A yellow or orange precipitate proves *cadmium* (9).

Notes

1. The HNO_3 used to dissolve the sulfides of the copper group must be dilute, since concentrated HNO_3 may oxidize PbS to $PbSO_4$.

$$PbS + 8\ HNO_3 \rightarrow PbSO_4 + 4\ H_2O + 8\ NO_2$$

$PbSO_4$ is insoluble and would be centrifuged off and thus lost from the solution.

2. HgS is insoluble in dilute HNO_3, and if any remains undissolved by Na_2S_2 in the separation of the copper and arsenic groups, it will appear as an insoluble residue in the copper group analysis.

3. The H_2SO_4 is evaporated to fumes of SO_3 to expel HNO_3 from the solution. When fumes of SO_3 appear, all HNO_3 has been expelled, and the H_2SO_4 has started decomposing.

$$H_2SO_4 \rightarrow H_2O + SO_3\uparrow$$

It is necessary to expel HNO_3, since (1) $PbSO_4$ is moderately soluble in HNO_3, and (2) later in the procedure when H_2S is used, HNO_3 would oxidize H_2S to free sulfur, which makes the analysis difficult.

4. $PbSO_4$ is soluble in hot ammonium acetate solution because slightly ionized $Pb(C_2H_3O_2)_2$ is formed. $PbCrO_4$ may then be precipitated from this solution by adding CrO_4^{2-}. (This is a good example of the application of comparative solubility products and ionization constants.)

$$K_{SP} \text{ for } PbSO_4 = 2.3 \times 10^{-8}$$

$$K_{SP} \text{ for } PbCrO_4 = 1.8 \times 10^{-14}$$

The ionization of $Pb(C_2H_3O_2)_2$ gives a Pb^{2+} concentration intermediate between $PbSO_4$ and $PbCrO_4$.

5. After the solution has been evaporated to fumes of SO_3, it is diluted to dissolve any sulfates of Bi, Cu, or Cd that may have precipitated in the concentrated acid solution.

6. Most of the Pb^{2+} is removed in Procedure 4 as $PbCl_2$; consequently, only small amounts of Pb^{2+} will be found in Procedure 6.

7. The test for Cu^{2+} with NH_3 is very sensitive, the blue color being visible where as little as 1 part of the $Cu(NH_3)_4^{2+}$ ion is present per 25,000 parts of H_2O.

8. Sodium stannite solution is prepared by adding an excess of NaOH to a solution of $SnCl_2$ (tin(II) chloride). The reaction takes place in two steps:

$$Sn^{2+} + 2\ OH^- \rightarrow Sn(OH)_2$$

$$Sn(OH)_2 + 2\ OH^- \rightarrow Sn(OH)_4^{2-} \quad \text{or} \quad SnO_2^{2-} + 2\ H_2O$$

The stannite solution must be made up just before using, since it oxidizes in the air to form a stannate solution that does not react with $Bi(OH)_3$.

9. CdS may precipitate in a colloidal condition, and the solution will appear yellow or orange. Addition of 1 or 2 drops of concentrated NH_3 will cause precipitation of CdS. A very pale yellow color is probably due to some decomposition products of thioacetamide.

Procedure 7 Analysis of Arsenic Subgroup

Solution: (From Procedure 5.) HgS_2^{2-}, AsS_4^{3-}, SbS_4^{3-}, SnS_3^{2-}. Add 6 *M* HCl dropwise. Stir after each drop until acid to litmus (1). Centrifuge and discard liquid (2).

Precipitate: As_2S_5, HgS, Sb_2S_5, SnS_2, S. Add 10 drops concentrated (12 *M*) HCl. Heat to nearly boiling with stirring. Centrifuge. Separate solution and treat precipitate with concentrated HCl as above and combine the two solutions. Treat the combined solutions with H_2S (2 drops thioacetamide for 5 minutes in hot water bath) to reprecipitate any HgS that may have dissolved. Add any precipitate to that already taken out.

Solution: Sb^{3+}, Sn^{4+}. Boil this solution gently until H_2S has been completely expelled.* Divide the solution into two parts, about ⅓ in part **(a)** and ⅔ in part **(b)**.

(a) To the solution add about 0.1 g oxalic acid solid; then add 8–10 drops H_2O. Warm and saturate with H_2S (use thioacetamide). Orange precipitate proves *antimony* (5).

(b) Add a short strip of Mg ribbon and then a few drops 6 *M* HCl. After reaction ceases, centrifuge if there is a residue, and add 1 drop $HgCl_2$ to the clear liquid. A white or grey precipitate proves *tin* (6).

Precipitate: HgS, As_2S_5. Add 10 drops 6 *M* NH_3. Stir for about 1 minute, then centrifuge. Separate solution from precipitate. Wash precipitate with 5 drops NH_3 (3).

Solution: AsS_4^{3-}. Acidify with 6 *M* HCl. Formation of yellow precipitate proves *arsenic* (4).

(continued)

*Use a piece of filter paper that has been moistened with $Pb(NO_3)_2$ solution. Place the paper in the fumes over the tube. No blackening will appear after H_2S is expelled.

Procedure 7 *(continued)*

Precipitate: HgS, S. Treat with 5 drops concentrated (12 *M*) HCl and 2 drops 6 *M* HNO_3. Transfer to a crucible and evaporate nearly to dryness. Add 3 drops H_2O. Centrifuge and filter if necessary to remove sulfur. Transfer solution to a test tube and add 2 drops $SnCl_2$. A white or grey precipitate proves *mercury*.

Notes

1. The acid used in reprecipitation of the arsenic group should be dilute since Sb_2S_5 and SnS_2 are soluble in moderately concentrated acid.

2. Addition of acid to a solution of Na_2S_2 results in formation of a white milky precipitate of sulfur.

$$Na_2S_2 + 2\ HCl \rightarrow 2\ NaCl + H_2S \uparrow + \underline{S}$$

Hence a precipitate will always be obtained when the solution of sodium polysulfide is acidified, whether members of the arsenic group are present or not. Sulfides of the arsenic group are colored and coagulate more readily than the free sulfur and so appear as a colored residue in the milky solution.

3. Test wash solution for complete extraction of the arsenic by acidifying a drop of the solution in a watch glass. No yellow precipitate will appear if the extraction is complete.

4. Arsenic can be confirmed as follows: Tear a piece of filter paper such that it can be rested and crease-clamped upon a 6-inch test tube. Place a small crystal of $AgNO_3$ on top of the paper. Place the solution to be tested for arsenic in the test tube. Add a few zinc granules (or some powdered zinc), then several drops of 6 *M* HCl to generate hydrogen gas. The latter forms AsH_3 with arsenic present. Arsine (AsH_3) gas rising in the tube and through the filter paper turns an $AgNO_3$ crystal yellow, then black.

Antimony forms stibine (SbH_3) if present, which turns the crystal black only. If sulfur is a contaminant, use concentrated NaOH instead of HCl to generate hydrogen gas with zinc, since H_2S will turn $AgNO_3$ black. $AgNO_3$ crystals should not be handled with the fingers as black stains may be produced on the skin.

5. If a large amount of Sn^{4+} is present, some SnS_2 may precipitate here. However it is yellow to tan in color in contrast to red-orange Sb_2S_3.

6. A test for tin can be made on the original sample. To 3–4 drops of original sample, add 6 *M* HCl until distinctly acid. Add 2 or 3 short strips of Mg ribbon. After reaction ceases, filter off any residue. To the clear solution add 2 drops $HgCl_2$. A white precipitate indicates tin.

Analysis of Unknown

When you have completed the analysis of the known solution of ions for this group, write the equations for the reactions of each ion. Then the instructor will issue an unknown containing ions of this group. *Use not more than 0.5 mL of the sample for analysis.* As you analyze the sample, prepare a diagram for the analysis similar to the one on page 44.

GROUP III: THE ALUMINUM–NICKEL–IRON GROUP

13

Chemistry of Separations and Tests

Precipitation of Group III and separation of subgroups

1. The basis for the separation of Group III from Groups IV and V is that sulfides and hydroxides of Group III precipitate in ammoniacal solution, whereas sulfides of Groups IV and V are soluble in this solution and hence do not precipitate. To prevent the precipitation of $Mg(OH)_2$, NH_4Cl is added to repress the ionization of NH_3 and reduce the OH^- concentration to a point where $Mg(OH)_2$ cannot form.

The separation of Group III from the moderately soluble hydroxides of Group IV and the slightly soluble $Mg(OH)_2$ of Group V depends on a carefully buffered solution. If NH_3 is added to an aqueous solution of $MgCl_2$ and such a representative member of Group III as $AlCl_3$, the hydroxides of both magnesium and aluminum will precipitate. Therefore, it is necessary to reduce the hydroxide concentration to a point where $Al(OH)_3$ will precipitate but $Mg(OH)_2$ will not. Consider the following:

$$K_{SP} \text{ of } Mg(OH)_2 = 3.4 \times 10^{-11}$$

$$K_{SP} \text{ of } Al(OH)_3 = 2 \times 10^{-33}$$

In 1 M NH_3, the concentration of OH^- is 4.24×10^{-3} M. Then

$$[Mg^{2+}] \times (4.24 \times 10^{-3})^2 = 3.4 \times 10^{-11}$$

$$[Mg^{2+}] = 1.9 \times 10^{-6} \text{ (minimum concentration to start precipitation)}$$

$$[Al^{3+}] \times (4.24 \times 10^{-3})^3 = 2 \times 10^{-33}$$

$$[Al^{3+}] = 2.6 \times 10^{-25} \text{ (minimum concentration to start precipitation)}$$

Now consider buffering the NH_3 solution with NH_4Cl so that the concentration of OH^- is decreased to 1.8×10^{-5}; then

$$[Mg^{2+}] = \frac{3.4 \times 10^{-11}}{(1.8 \times 10^{-5})^2} = 0.11$$

This concentration (0.11) of Mg^{2+} is above that normally present in solutions for analysis; therefore, $Mg(OH)_2$ does not precipitate in Group III. However, $Al(OH)_3$ and other hydroxides of Group III do precipitate. For example,

$$[Al^{3+}] = \frac{2 \times 10^{-33}}{(1.8 \times 10^{-5})^3} = 3.4 \times 10^{-19}$$

Above this concentration of Al^{3+}, $Al(OH)_3$ will precipitate—even in buffered NH_3.

Group III: The Aluminum–Nickel–Iron Group

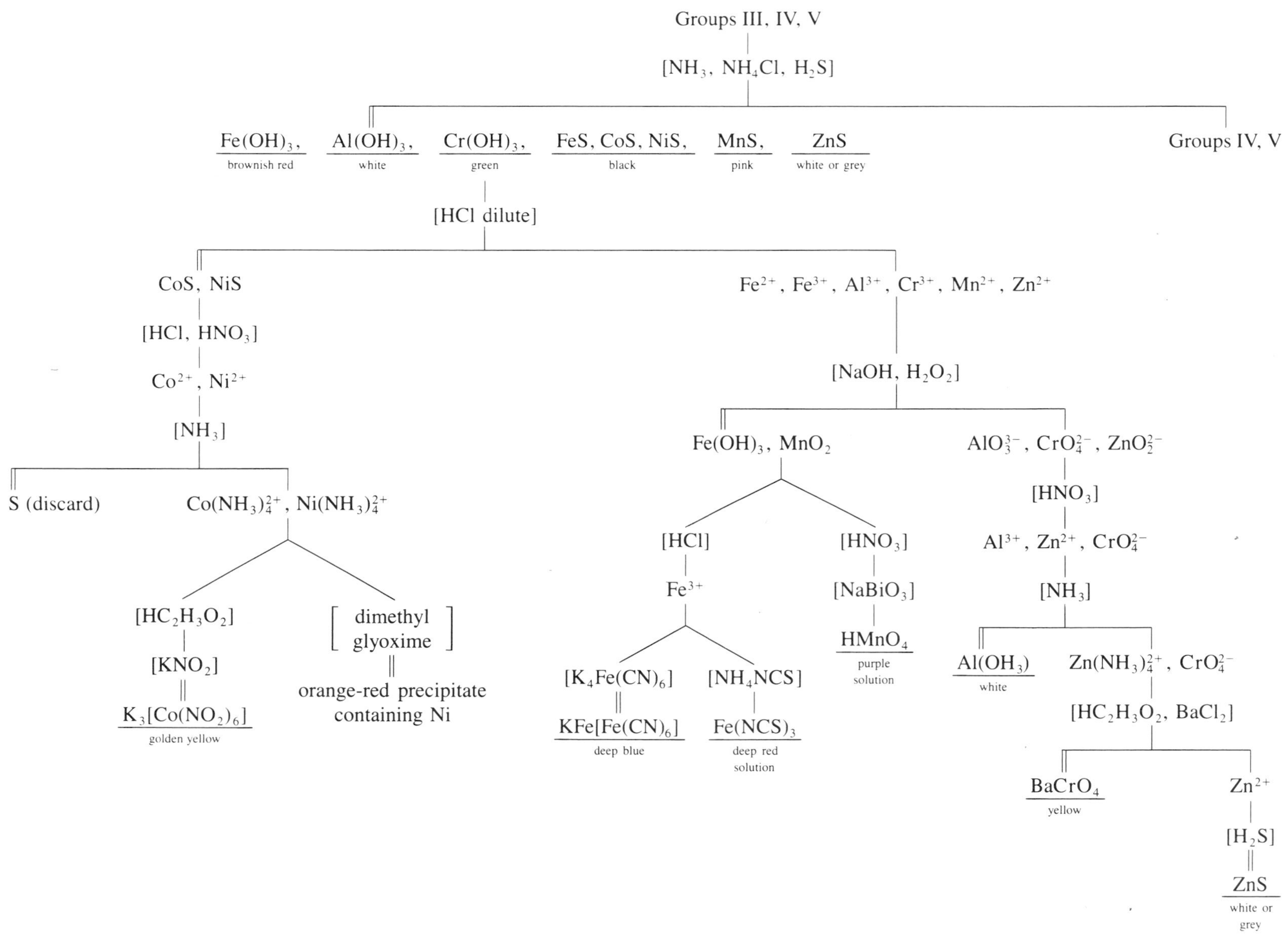

2. Several of the ions in this group are colored, and often some information regarding the presence of cation members of Group III in the sample may be gained by observing the color of the solution. Colors of ions are: Co^{2+} pink; Ni^{2+} green; Fe^{2+} very pale green; Cr^{3+} dark green; Mn^{2+} pale pink to colorless. Fe^{3+} in the presence of Cl^- forms a complex ion that is yellow. Obviously, if the solution to be analyzed for Group III appears colorless, those ions with distinct colors (Ni^{2+}, Cr^{3+}, Co^{2+}) can be eliminated as possibilities.

3. Additional information can be obtained after the solution is made ammoniacal with NH_3. Zinc, nickel, and cobalt form complex ions and thus are not precipitated until after addition of H_2S. However, iron, chromium, and aluminum precipitate as hydroxides: $Fe(OH)_3$ red-brown; $Cr(OH)_3$ green; $Al(OH)_3$ white. If no precipitate forms on making the solution ammoniacal, then Fe^{3+}, Cr^{3+}, Al^{3+} can be eliminated as possibilities. Reactions of the ions with NH_3 are

$$Ni^{2+} + 4\,NH_3 \rightarrow Ni(NH_3)_4^{2+}$$

$$Co^{2+} + 4\,NH_3 \rightarrow Co(NH_3)_4^{2+}$$

$$Zn^{2+} + 4\,NH_3 \rightarrow Zn(NH_3)_4^{2+}$$

$$Fe^{3+} + 3\,NH_3 + 3\,H_2O \rightarrow \underline{Fe(OH)_3} + 3\,NH_4^+$$

$$Al^{3+} + 3\,NH_3 + 3\,H_2O \rightarrow \underline{Al(OH)_3} + 3\,NH_4^+$$

$$Cr^{3+} + 3\,NH_3 + 3\,H_2O \rightarrow \underline{Cr(OH)_3} + 3\,NH_4^+$$

4. On addition of H_2S, the three insoluble hydroxides above remain as such, but the other ions produce insoluble sulfides.

$$Ni(NH_3)_4^{2+} + S^{2-} \rightarrow \underline{NiS} + 4\,NH_3$$

$$Fe^{2+} + S^{2-} \rightarrow \underline{FeS}$$

Some $Fe(OH)_3$ may be converted to FeS.

$$2\,Fe(OH)_3 + 3\,H_2S \rightarrow \underline{2\,FeS} + \underline{S} + 6\,H_2O$$

5. Co and Ni can be separated from other members of the group because NiS and CoS are very slowly soluble in dilute HCl, whereas the sulfides and hydroxides of the other metals dissolve readily. In effecting the separation it is very important, then, that NiS and CoS not remain in contact with HCl very long.

6. Iron and manganese can be separated from aluminum, chromium, and zinc because the hydroxides of the latter three elements are amphoteric; that is, their hydroxides are soluble in excess strong base, such as NaOH.

$$\underline{Zn(OH)_2} + 2\,OH^- \rightarrow Zn(OH)_4^{2-} \quad \text{or} \quad ZnO_2^{2-} + 2\,H_2O$$

$$\underline{Al(OH)_3} + 3\,OH^- \rightarrow Al(OH)_6^{3-} \quad \text{or} \quad AlO_3^{3-} + 3\,H_2O$$

$$\underline{Al(OH)_3} + OH^- \rightarrow Al(OH)_4^- \quad \text{or} \quad AlO_2^- + 2\,H_2O$$

$$\underline{Cr(OH)_3} + 3\,OH^- \rightarrow Cr(OH)_6^{3-} \quad \text{or} \quad CrO_3^{3-} + 3\,H_2O$$

Hydrogen peroxide is added to convert $Cr(OH)_6^{3-}$ to CrO_4^{2-}, so that aluminum can be effectively separated from chromium.

$$2\,Cr(OH)_6^{3-} + 3\,H_2O_2 \rightarrow 2\,CrO_4^{2-} + 8\,H_2O + 2\,OH^-$$

$Mn(OH)_2$ and $Fe(OH)_2$ are also oxidized by H_2O_2.

$$\underline{Mn(OH)_2} + H_2O_2 \rightarrow \underline{MnO_2} + 2\,H_2O$$

$$\underline{2\,Fe(OH)_2} + H_2O_2 \rightarrow \underline{2\,Fe(OH)_3}$$

The above separations yield three subgroups: nickel subgroup (Co, Ni); iron subgroup (Fe, Mn); aluminum subgroup (Al, Cr, Zn).

Analysis of nickel subgroup

7. CoS and NiS are most readily dissolved by aqua regia.

$$CoS + 4\,H^+ + Cl^- + NO_3^- \rightarrow Co^{2+} + S + NOCl + 2\,H_2O$$

Tests for Co and Ni may be made in the presence of one another; consequently, it is unnecessary to effect a separation. Dimethyl glyoxime produces a bright red precipitate with Ni^{2+} in ammoniacal solution, and potassium nitrite produces a golden yellow precipitate in acetic acid solution with cobalt, which is air–oxidized to Co^{3+} in the process.

Analysis of iron subgroup

8. Iron and manganese ions can be tested for in the presence of each other by the following: Fe^{3+} is readily tested for by addition of $K_4Fe(CN)_6$, which gives a dark blue precipitate of $KFe[Fe(CN)_6]$, called Prussian blue.

$$Fe^{3+} + K^+ + Fe(CN)_6^{4-} \rightarrow \underline{KFe[Fe(CN)_6]}$$

An alternate test is addition of NCS^- (NH_4NCS), which forms a red thiocyanate complex with Fe^{3+}, probably $FeNCS^{2+}$.

9. The test for Mn is based on the oxidation of MnO_2 to violet-colored permanganate ion, MnO_4^-. This is readily accomplished with PbO_2 or $NaBiO_3$ (sodium bismuthate) and HNO_3. HCl or Cl^- should be absent.

$$2\,MnO_2 + 3\,PbO_2 + 4\,H^+ \rightarrow 2\,MnO_4^- + 3\,Pb^{2+} + 2\,H_2O$$

$$2\,MnO_2 + 3\,NaBiO_3 + 10\,H^+ \rightarrow 2\,MnO_4^- + 3\,Na^+ + 3\,Bi^{3+} + 5\,H_2O$$

10. The basis for the separation of Al^{3+} from CrO_4^{2-} and Zn^{2+} is the reaction of NH_3 with Al^{3+} to form $Al(OH)_3$, which precipitates, and the reaction of NH_3 with Zn^{2+} to form a complex ion, namely $Zn(NH_3)_4^{2+}$, that remains in solution. CrO_4^{2-} is unaffected by NH_3.

$$Al^{3+} + 3\,NH_3 + 3\,H_2O \rightarrow \underline{Al(OH)_3} + 3\,NH_4^+$$

$$Zn^{2+} + 4\,NH_3 \rightarrow Zn(NH_3)_4^{2+}$$

The white flocculent precipitate of $Al(OH)_3$ on addition of NH_3 indicates Al^{3+}. Very concentrated NH_3 will redissolve $Al(OH)_3$.

$$Al(OH)_3 + 3\,OH^- \rightarrow Al(OH)_6^{3-}$$

Certain organic substances are adsorbed by $Al(OH)_3$ to give a colored product, and these are sometimes helpful in confirming its presence. Aurin tricarboxylic acid, called *aluminon*, gives a red color to the $Al(OH)_3$; similarly, arsenazo agent gives a purple color.

11. The separation of Zn^{2+} and CrO_4^{2-} depends on the fact that Cr is a part of the anion CrO_4^{2-} and can be precipitated as $BaCrO_4$ in acetic acid solution on addition of Ba^{2+}.

$$CrO_4^{2-} + Ba^{2+} \rightarrow \underline{BaCrO_4}$$

Zn^{2+} then can be precipitated in the weakly acid filtrate.

$$Zn^{2+} + H_2S \rightarrow \underline{ZnS} + 2\,H^+$$

Zinc can be confirmed by dissolving the ZnS precipitate in HCl, heating to remove H_2S, making the solution basic with NaOH, and adding a drop to dithizone test paper; a red-purple color confirms Zn^{2+}. An alternate test is the following:

$$3\ Zn^{2+} + 2\ K^{+} + 2\ Fe(CN)_6^{4-} \rightarrow Zn_3K_2[Fe(CN)_6]_2$$

Zinc can also be confirmed by treating the ZnS precipitate with HNO_3 and $Co(NO_3)_2$ and heating the mixture to a high temperature in a crucible; if the product is green, zinc is assumed present.

$$CoO + ZnO \rightarrow CoZnO_2 \text{ (Rinmann's green)}$$

Preliminary Tests on Group III

Record your observations and write the corresponding chemical equations.

1. Effect of NH_3 and NH_4Cl on Group III cations. Place 1 drop of test solution of each ion of Group III (Fe^{3+}, Al^{3+}, Mn^{2+}, Cr^{3+}, Ni^{2+}, Co^{2+}, Zn^{2+}) into separate test tubes. Dilute each to about 0.5 mL, and add 1 drop of dilute NH_3 to each. Write the equations. Now add 3 drops of concentrated NH_3 to each tube. Note which precipitates, if any, dissolve. Repeat the tests on Ni^{2+}, Co^{2+}, and Zn^{2+}, first adding 2 drops of NH_4Cl solution.

2. Action of NaOH and H_2O_2 on $Cr(OH)_3$. Place 1 drop of Cr^{3+} test solution in a test tube and dilute to 0.5 mL. Add 1 drop NaOH. Centrifuge the tube containing $Cr(OH)_3$, and discard the solution. To the precipitate add 2 drops of NaOH and a few drops of H_2O_2, and warm.

3. Effect of NaOH on aluminum and zinc ions. To several drops of test solution of Al^{3+} and Zn^{2+} in separate test tubes add 1 drop of NaOH and observe. Then add an excess of NaOH to each.

4. A means of preventing $Mg(OH)_2$ from precipitating in Group III. To each of two test tubes add 1 drop of Mg^{2+} test solution (Mg is in Group V). To one tube add a drop of NH_4Cl solution. Now add NH_3 to each tube. (See theoretical discussion on page 53.)

Analytical Procedures for Group III

Make up a known solution using 3 drops of test solution of each ion of this group. Analyze according to Procedures 8–11.

Procedure 8 Precipitation and Separation of Subgroups of Group III

Solution: (From Procedure 5, containing cations of Groups III, IV, V.) (1) Adjust the volume of solution to about 1.5 mL; add 2 drops NH_4Cl solution (2); then add 6 *M* NH_3 until just basic. Add 2 drops 15 *M* NH_3; then saturate with H_2S (3), (4), using thioacetamide. Centrifuge.

(continued)

Procedure 8 *(continued)*
Solution: Add 6 *M* acetic acid until acid; evaporate to 1 mL (5). Centrifuge and discard any precipitate. Decant supernatant liquid and test for Groups IV and V by Procedure 12.
Precipitate: CoS (black), NiS (black), FeS (black), $Al(OH)_3$ (white), $Cr(OH)_3$ (green), MnS (pink), ZnS (white), $Fe(OH)_3$ (red-brown). Wash twice with H_2O. Discard washings. To the precipitate in the test tube add 12 drops H_2O and 2 drops concentrated (12 *M*) HCl. Stir thoroughly for about 1 minute. Centrifuge immediately.
Precipitate: CoS, NiS. Treat according to Procedure 9.
Solution: Fe^{2+}, Fe^{3+}, Mn^{2+}, Cr^{3+} (green), Al^{3+}, Zn^{2+}. Transfer to a crucible or small casserole; add NaOH until basic, then 4 drops more. Add 6–8 drops of 3% H_2O_2 slowly (6). Heat solution to near boiling for a few minutes and until the gas (oxygen) evolution ceases. Add 3 mL H_2O. Transfer to test tube and centrifuge.
Precipitate: Iron subgroup: $Fe(OH)_3$ (brownish red), MnO_2 (brown to black). Wash precipitate twice with hot H_2O. Discard washings. Treat precipitate according to Procedure 10.
Solution: Aluminum subgroup: AlO_3^{3-}, CrO_4^{2-} (yellow), ZnO_2^{2-}. Treat according to Procedure 11.

Notes

1. H_2S is removed before addition of NH_3 to gain information about precipitation with NH_3 alone. (See Procedure 5.) Any Group II sulfides that might be in the filtrate in a colloidal form are precipitated when H_2S is expelled. These may be centrifuged and discarded.

2. NH_4Cl is added to prevent precipitation of $Mg(OH)_2$ and to promote coagulation of sulfides, which tend to precipitate in colloidal form.

3. When H_2S is added to an ammoniacal solution, a high concentration of S^{2-} results. S^{2-} thus may react directly with cations to produce insoluble sulfides.

4. It is important that the solution not be too ammoniacal when H_2S is added. Excess NH_3 tends to hold certain sulfides, notably CoS and NiS, in colloidal condition.

5. The filtrate of Groups IV and V is acidified with $HC_2H_3O_2$ and evaporated for removal of colloidal constituents that may have remained in suspension.

6. Sodium peroxide may be used here instead of H_2O_2. The former is added slowly from the tip of a spatula. (Any unused Na_2O_2 should be disposed of in the sink, since combustible material such as paper may be ignited because of the strong oxidizing character of the peroxide. Also, on exposure to air, Na_2O_2 becomes sticky and hard from the formation of NaOH and Na_2CO_3 from moisture and CO_2 in the air.)

Procedure 9 Analysis of Nickel Subgroup

Precipitate: (From Procedure 8.) Wash twice with cold H_2O and reject the washings. To the precipitate add 5 drops 6 *M* HCl and 1 drop 6 *M* HNO_3. Heat until precipitate dissolves. Add 5–6 drops H_2O. Add 15 *M* NH_3 dropwise to the solution until basic, then 1 drop more. Centrifuge.

Precipitate: Some S from oxidation of H_2S. Discard.

Solution: $Co(NH_3)_4^{2+}$, $Ni(NH_3)_4^{2+}$. Divide the filtrate into two equal parts, **(a)** and **(b)**.

(a) Add acetic acid until acid, then 4 drops of saturated KNO_2 solution. Warm and stir for a few minutes, then let cool for 15 minutes more (1). Formation of a golden yellow precipitate proves *cobalt*.

(b) Add 3 drops dimethyl glyoxime and 1 drop 15 *M* NH_3. Red precipitate proves *nickel*.

Notes

1. The precipitate of cobalt with KNO_2 forms very slowly and requires the presence of air. Hence, the solution should be allowed to stand 10–15 minutes before any conclusion is drawn about the absence of cobalt.

Procedure 10 Analysis of Iron Subgroup

Precipitate: (From Procedure 8.) Divide precipitate into two parts; about ⅔ in part **(a)** and ⅓ in part **(b)**.

(a) To part of the precipitate add 5 drops 6 *M* HNO_3 and 5 drops H_2O. Add a pinch of $NaBiO_3$ and mix thoroughly. Heat and centrifuge. A violet supernatant liquid proves *manganese*. (PbO_2 may be used in place of $NaBiO_3$. Cl^- should not be present in making this test.)

(b) Add 5 drops 6 *M* HCl and heat until dissolved. Divide into two parts. Cool and add 1 drop $K_4Fe(CN)_6$ (1),(2) to one part. Formation of a deep blue precipitate of Prussian blue proves *iron*. To the second portion add 1 drop NH_4NCS. A deep red solution verifies *iron*.

Notes

1. Whether originally present as Fe^{2+} or Fe^{3+}, iron will appear at this point in the procedure as Fe^{3+}. To determine the state of oxidation of iron in the original sample, solutions of potassium ferrocyanide and potassium ferricyanide may be used. Dissolve a

small portion of the original sample in dilute HCl. Divide into two parts; to the first part add a few drops of $K_4Fe(CN)_6$. A deep Prussian blue precipitate proves iron(III) ion. To the second part add a few drops of $K_3Fe(CN)_6$. A precipitate of deep Prussian blue proves presence of iron(II) ion. The same blue precipitate is obtained in both tests.

$$Fe^{3+} + K^+ + Fe(CN)_6^{4-} \rightarrow \underset{\text{Prussian blue}}{\underline{KFe[Fe(CN)_6]}}$$

$$Fe^{2+} + K^+ + Fe(CN)_6^{3-} \rightarrow \underset{\text{Prussian blue}}{\underline{KFe[Fe(CN)_6]}}$$

If a deep blue precipitate is obtained with both ferro- and ferricyanide solutions, then both iron(II) and iron(III) ions are present. A solution containing Fe^{3+} forms a deep red solution of $FeNCS^{2+}$ with NCS^-.

2. A light blue color in the test for iron is probably due to a trace of iron or some other metallic ion that has not been removed completely. If the test does not give a *deep, dark blue* precipitate, iron may be assumed to be absent.

Procedure 11 Analysis of Aluminum Subgroup

Solution: (From Procedure 8.) Add 6 *M* HCl in a small evaporating dish until just acid (1); then add 15 *M* NH_3 dropwise until basic. Add 3 more drops NH_3. Heat to boiling. Transfer to test tube. Formation of white flocculent precipitate proves *aluminum*. Centrifuge.

Precipitate: $Al(OH)_3$ (white) (7).

Solution: CrO_4^{2-} (yellow), $Zn(NH_3)_4^{2+}$. If the solution is yellow, *chromium* is present (2). Acidify with acetic acid (3). Add a few crystals of $NaC_2H_3O_2$, and then 2 drops $BaCl_2$. Formation of a yellow precipitate confirms *chromium*. Centrifuge.

Precipitate: $BaCrO_4$ (yellow) and $BaSO_4$ (white).

Solution: Divide solution into two parts. Add 3–4 drops of 0.1 *M* $K_4Fe(CN)_6$ to one part for slightly grey $Zn_3K_2[Fe(CN)_6]_2$ precipitate. Saturate other portion with H_2S using thioacetamide. A grey-white precipitate is probably ZnS (4). To confirm, centrifuge and wash precipitate twice with H_2O. Discard washings. Add 6–8 drops 6 *M* HCl to residue. If precipitate dissolves, presence of *zinc* is confirmed (5),(6).

Notes

1. Chromium chemistry involves several oxidation states. Cr^{3+} treated with excess NaOH forms CrO_3^{3-} [$Cr(OH)_6^{3-}$, chromite], much as Al^{3+} treated with excess NaOH forms AlO_3^{3-} [$Al(OH)_6^{3-}$, aluminate]. Addition of acid (HCl or HNO_3) converts back to Cr^{3+} and Al^{3+}. If CrO_3^{3-} has not been oxidized to yellow stable CrO_4^{2-}, Cr^{3+} will follow Al^{3+} and a mixed precipitate of $Al(OH)_3$ and $Cr(OH)_3$ forms. The $Cr(OH)_3$

would thus interfere with the final $Al(OH)_3$ test. $Cr(OH)_3$ in small amounts is light green and may be mistaken for $Al(OH)_3$.

The reason for 6 *M* HCl instead of 6 *M* HNO_3 in the treatment of AlO_3^{3-}, ZnO_2^{2-}, and CrO_4^{2-} to convert to Al^{3+}, Zn^{2+}, and CrO_4^{2-} is because CrO_4^{2-} in nitrate-containing solution and with residual H_2O_2 may yield an indigo-blue peroxide compound to which the formula CrO_5 has been given, and which may decompose to Cr^{3+}.

$$2\ CrO_4^{2-} + 2\ H^+ \rightleftharpoons Cr_2O_7^{2-} + H_2O$$

$$2\ H^+ + Cr_2O_7^{2-} + 4\ H_2O_2 \rightleftharpoons 2\ CrO_5 + 5\ H_2O$$

The unstable colored CrO_5 more or less rapidly decomposes with oxygen evolution (decomposition rate is dependent on HNO_3 and H_2O_2 concentration) to Cr^{3+} ion.

$$4\ CrO_5 + 12\ H^+ \rightarrow 4\ Cr^{3+} + 6\ H_2O + 7\ O_2$$

The indigo-blue CrO_5 is soluble and more stable in ether. Thus a verifying test for yellow CrO_4^{2-} is to use a $BaCrO_4$ precipitate suspended in 2 mL H_2O or 2 mL yellow CrO_4^{2-} solution in a test tube. Add 5 drops 3 *M* HNO_3 and heat to boiling. Then cool under tap water. Add 1 mL ethyl ether to float above the CrO_4^{2-} solution and 2 drops 3% H_2O_2. Stir. A blue color in the ether layer confirms chromium.

2. If CrO_4^{2-} is present, the filtrate from Procedure 8 should be yellow. The yellow color is a sensitive test for CrO_4^{2-}, and if solution from Procedure 8 is colorless, assume that chromium is absent and go directly to the test for zinc.

3. Since $BaCrO_4$ is soluble in HCl, HNO_3, and other strong acids, the precipitation of $BaCrO_4$ is made in $HC_2H_3O_2$ in which it is insoluble. The use of $NaC_2H_3O_2$ is to yield ions to help coagulate $BaCrO_4$, which tends to be colloidal. A few minutes heating in a water bath should yield $BaCrO_4$ precipitate. See also note 1 for an additional confirmatory test for chromium.

4. In precipitation of ZnS with thioacetamide, some sulfur may precipitate in colloidal form; this white precipitate may be mistaken for white ZnS. Both ZnS and $Zn_3K_2[Fe(CN)_6]_2$ are insoluble in slightly acid solution. Here the solution from the above manipulation should be slightly acid with acetic acid.

5. Zinc may also be confirmed by adding excess 5% NaOH to a few drops of the original sample and testing with dithizone test paper. A red-violet color to the test paper is produced if Zn^{2+} is present. NaOH gives an orange color that must not be mistaken for the red-violet color with Zn^{2+}. A known sample of Zn^{2+} should always be used side by side with the unknown when using the dithizone test paper.

6. Zinc may be confirmed by making a paste of precipitated ZnS with 2 drops 0.05 *M* $Co(NO_3)_2$ and 1 drop dilute HNO_3. [A very small crystal of $Co(NO_3)_2$ may be used instead of 2 drops.] Place paste in a small crucible (or evaporating dish). Dry carefully and heat the crucible bottom to redness in the hottest part of the Bunsen flame (high temperature essential). Formation of green $CoZnO_2$ confirms zinc. Colored glaze on porcelain is made commercially in this manner. Here, the green glaze can be wiped off.

7. To confirm Al^{3+}, suspend precipitate in 10 drops H_2O; then add dilute acetic acid until solution is slightly acid. Add 4 drops aluminon reagent, and heat in water bath. When it is heated add a few drops of dilute NH_3 until it is neutral or slightly basic. Red precipitate confirms Al^{3+}. In all tests using organic reagents, run a known sample along with the unknown for a comparison. The arsenazo reagent test for Al is made as follows: Dissolve some $Al(OH)_3$ precipitate in dilute $HC_2H_3O_2$ ($pH = 3-7$). Add 2 drops arsenazo reagent (0.001 *M*). If Al is present, a reddish purple color results; if Al is absent, an orange color results.

Analysis of Unknown Solution

When the analysis of the known solution has been completed, write equations for all reactions of ions of this group. Then you will receive an unknown containing ions in this group. Use 0.5 mL of the sample for analysis. Analyze according to Procedures 8–11, developing a diagram of analysis similar to the one on page 54 as you progress.

GROUP IV: THE BARIUM GROUP

14

Chemistry of Separations and Tests

Separation of barium and alkali groups and analysis of the barium group

1. The addition of $(NH_4)_2CO_3$ to an alkaline solution of the Ba and alkali groups in the presence of excess NH_4^+ (using NH_4Cl) results in the precipitation of $BaCO_3$, $SrCO_3$, and $CaCO_3$. Mg^{2+}, K^+, Na^+, and NH_4^+ remain in solution.

$$Ba^{2+} + CO_3^{2-} \rightarrow \underline{BaCO_3}$$

$$Sr^{2+} + CO_3^{2-} \rightarrow \underline{SrCO_3}$$

$$Ca^{2+} + CO_3^{2-} \rightarrow \underline{CaCO_3}$$

2. The separation of Ba^{2+} and Sr^{2+} from Ca^{2+} is based on the slight solubility of the nitrates of barium and strontium in absolute (also called dry or 100%) alcohol; $Ca(NO_3)_2$ is soluble. For example, the solubilities at 25°C in mg of salt per mL of absolute ethyl alcohol are approximately

$Ca(NO_3)_2$	$Ba(NO_3)_2$	$Sr(NO_3)_2$
850	0.02	0.06

Dry isopropyl alcohol or acetone may be used in place of ethyl alcohol.

3. The separation of Ba^{2+} from Sr^{2+} is based on the fact that $BaCrO_4$ is insoluble in acetic acid solution, whereas $SrCrO_4$ is soluble. Thus if chromate ion (K_2CrO_4) is added to an acetic acid solution of Ba^{2+} and Sr^{2+}, only the Ba^{2+} precipitates as $BaCrO_4$.

$$Ba^{2+} + CrO_4^{2-} \rightarrow \underline{BaCrO_4}$$

On addition of alcohol and OH^- to the solution, the solubility of $SrCrO_4$ is decreased, and it precipitates.

4. Ca^{2+} is precipitated with oxalate ion, $C_2O_4^{2-}$ in ammoniacal solution:

$$Ca^{2+} + C_2O_4^{2-} \rightarrow \underline{CaC_2O_4}$$

Preliminary Tests for Group IV

There is no necessity for preliminary tests for this group. Begin with the analytical procedures.

Group IV: The Barium Group

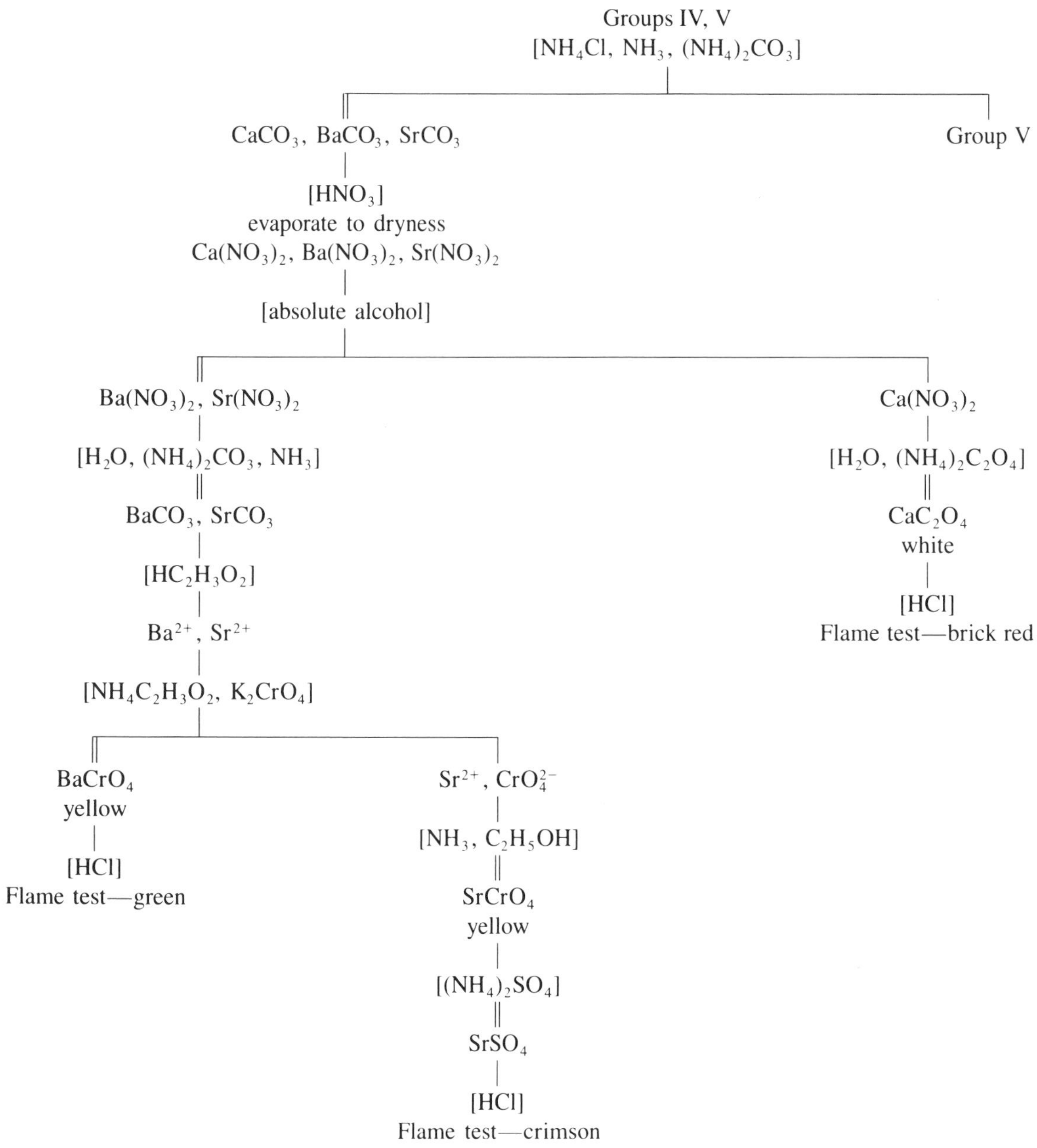

Analytical Procedures for Group IV

Make up a known sample of ions of this group by using 4 drops each of test solutions of the ions. Analyze according to Procedure 12. Write equations for reactions of the ions as you proceed.

Procedure 12 Separation of Barium and Alkali Groups and Analysis of Barium Group

Solution: (From Procedure 8.) If NH_4^+ is not already present, add 2 drops NH_4Cl (1). Add 6 *M* NH_3 dropwise until the solution is basic. Add 5 drops $(NH_4)_2CO_3$ (2), and place the tube in a hot water bath for a few minutes. Stir occasionally. Centrifuge. Add 1 drop more of $(NH_4)_2CO_3$ to the supernatant liquid to test for complete precipitation. Decant the solution.

Solution: Group V ions. Treat according to Procedure 13.

Precipitate: $BaCO_3$, $SrCO_3$, $CaCO_3$ (all white). Wash precipitate with a few drops of H_2O to which 1 drop $(NH_4)_2CO_3$ has been added. Discard the washings. Dissolve precipitate using 8 drops H_2O and 2 drops 6 *M* HNO_3. Evaporate to dryness in an evaporating dish, but do not bake. Stir and digest the residue with 2 mL absolute ethyl alcohol or isopropyl alcohol. Add more alcohol if necessary to keep the volume about 2 mL. Transfer solution and residue to a small test tube. Centrifuge and decant the solution (3).

Solution: $Ca(NO_3)_2(Mg^{2+})$. Add 2 mL H_2O. Evaporate to 1 mL and discard any precipitate. Add 2 drops dilute NH_3 and 2 drops $(NH_4)_2C_2O_4$. A white precipitate indicates calcium. Warm 2 minutes. Centrifuge. Save solution for Mg^{2+} test below. Treat the precipitate with 2 drops dilute HCl, and make a flame test. A brick-red flame test confirms calcium. Calcium flame test is not as definite as for Ba^{2+} or Sr^{2+}. Test the solution for Mg^{2+} by adding 2 drops Na_2HPO_4, and allow to stand for 5 minutes. A white crystalline precipitate indicates Mg^{2+}. If Mg^{2+} is found here, it should also be found in Group V.

Residue: $Ba(NO_3)_2$, $Sr(NO_3)_2$. Add 1 mL H_2O to dissolve, then reprecipitate $BaCO_3$, $SrCO_3$ with $(NH_4)_2CO_3$ as above. Centrifuge. Discard liquid. Dissolve precipitate in 3–4 drops 6 *M* acetic acid. Add 2 drops $NH_4C_2H_3O_2$ and 5 drops H_2O. Add 2 drops K_2CrO_4 (4). Stir for about 1 minute. A yellow precipitate indicates *barium*. Centrifuge.

Precipitate: $BaCrO_4$ (yellow). Wash the precipitate with H_2O until washings are colorless. Discard washings. Add 2 drops 6 *M* HCl. Dip a clean platinum (or nichrome) wire in the solution, and then put the wire in a nonluminous flame. A green flame test confirms *barium*.

Solution: Sr^{2+}, CrO_4^{2-}. Add 6 *M* NH_3 dropwise until basic to litmus. Concentrate the solution to 0.5–1 mL. Add an equal volume of ethyl alcohol while stirring the mixture (5). Cool under the tap and let stand 5 minutes. A yellow precipitate indicates *strontium*. Centrifuge. Do not wash precipitate.

Solution: Discard.

Precipitate: $SrCrO_4$. Add 10 drops $(NH_4)_2SO_4$ solution, and heat to boiling for about 1 minute in a water bath. Cool under the tap. Centrifuge. Wash precipitate of $SrSO_4$ with H_2O. Discard washings. Treat precipitate with 2 drops 6 *M* HCl, and make a flame test. A crimson color confirms *strontium* (6).

Notes

1. The concentration of OH^- is kept low by the presence of NH_4^+. Without this, the $[OH^-]$ would be so large that the product of $[Mg^{2+}] \times [OH^-]^2$ would exceed the solubility product and thus precipitate $Mg(OH)_2$.

2. $MgCO_3$ $K_{SP} = 2.6 \times 10^{-5}$

$BaCO_3$ $K_{SP} = 4.9 \times 10^{-9}$

$CaCO_3$ $K_{SP} = 1.7 \times 10^{-8}$

$SrCO_3$ $K_{SP} = 4.6 \times 10^{-9}$

If the concentration of Mg^{2+} in the original sample is high, some $MgCO_3$ will probably precipitate here. It probably will not interfere with the tests for Ba^{2+}, Sr^{2+}, and Ca^{2+}, however. It should be tested for in this group as well as in Group V. The concentration of CO_3^{2-} should be so regulated that all the $BaCO_3$ will be precipitated but $MgCO_3$ will not be—that is, $[Ba^{2+}] \times [CO_3^{2-}]$ should exceed its solubility product of 4.9×10^{-9}, but $[Mg^{2+}] \times [CO_3^{2-}]$ should not exceed its solubility product of 2.6×10^{-5}. Control of the hydroxide ion concentration aids in this regard. $(NH_4)_2CO_3$ is so extensively hydrolyzed that a solution of it (in the absence of NH_3) does not contain enough carbonate ions to completely precipitate $BaCO_3$, $CaCO_3$, and $SrCO_3$.

$$CO_3^{2-} + H_2O \rightleftharpoons HCO_3^- + OH^-$$

NH_3 is added, increasing the OH^- concentration to shift this equilibrium to the left. Since too great a concentration of OH^- must also be prevented (see note 1), the presence of NH_4^+ from some salt is essential.

3. Dry acetone or isopropyl alcohol may be used instead of dry (absolute) ethyl alcohol. These are solvents for $Ca(NO_3)_2$; $Ba(NO_3)_2$ and $Sr(NO_3)_2$ are relatively insoluble. H_2O in the solvent would dissolve all three nitrates, of course, and separation would not be effected. Solubilities compared:

$Ca(NO_3)_2$ in dry ethyl alcohol	85.4 g/100 g
$Sr(NO_3)_2$ in dry ethyl alcohol	0.0062 g/100 g
$Ba(NO_3)_2$ in dry ethyl alcohol	0.0016 g/100 g
$Ca(NO_3)_2$ in dry isopropyl alcohol	2.6 g/100 g
$Sr(NO_3)_2$ in dry isopropyl alcohol	0.002 g/100 g
$Ba(NO_3)_2$ in dry isopropyl alcohol	0.0016 g/100 g

4. The method of separating Ba^{2+} and Sr^{2+} is based on the relative solubilities of the chromates given below.

$BaCrO_4$ $K_{SP} = 2.3 \times 10^{-10}$

$SrCrO_4$ $K_{SP} = 3.8 \times 10^{-4}$

$CaCrO_4$ $K_{SP} = 2.3 \times 10^{-2}$

The CrO_4^{2-} ion concentration is to be controlled so that $BaCrO_4$ is precipitated, but strontium chromate is left in solution. The CrO_4^{2-} ion concentration may be controlled by the H^+ ion concentration, as shown by the equation

$$2\ CrO_4^{2-} + 2\ H^+ \rightleftharpoons Cr_2O_7^{2-} + H_2O$$

If the concentration of H^+ is too small, the concentration of CrO_4^{2-} will be high enough to precipitate $SrCrO_4$. If the concentration of H^+ increases, the concentration of CrO_4^{2-} decreases, so that $BaCrO_4$ may not precipitate completely. The proper concentration of H^+ is maintained in an acetic acid–ammonium acetate buffer solution. Thus the acetate ions hold H^+ and CrO_4^{2-} ions in reserve in the form of $HC_2H_3O_2$ and $Cr_2O_7^{2-}$, respectively.

5. $SrCrO_4$, although more soluble than $BaCrO_4$, will precipitate if the H^+ is low (that is, chromate high), and if alcohol is added. (Practically all inorganic salts are less soluble in alcohol than in H_2O.) In this reaction more K_2CrO_4 reagent need not be added, since the excess is still present from the original addition of it.

6. If $BaCrO_4$ was not entirely separated from the strontium and calcium salts, the flame test at this point will show a flash of green before the more persistent crimson color of strontium.

Analysis of Unknown Solutions

After studying Group V, run a mixed unknown of Group IV and V. No separate unknown of this three-ion group is necessary.

Group V: The Alkali Group

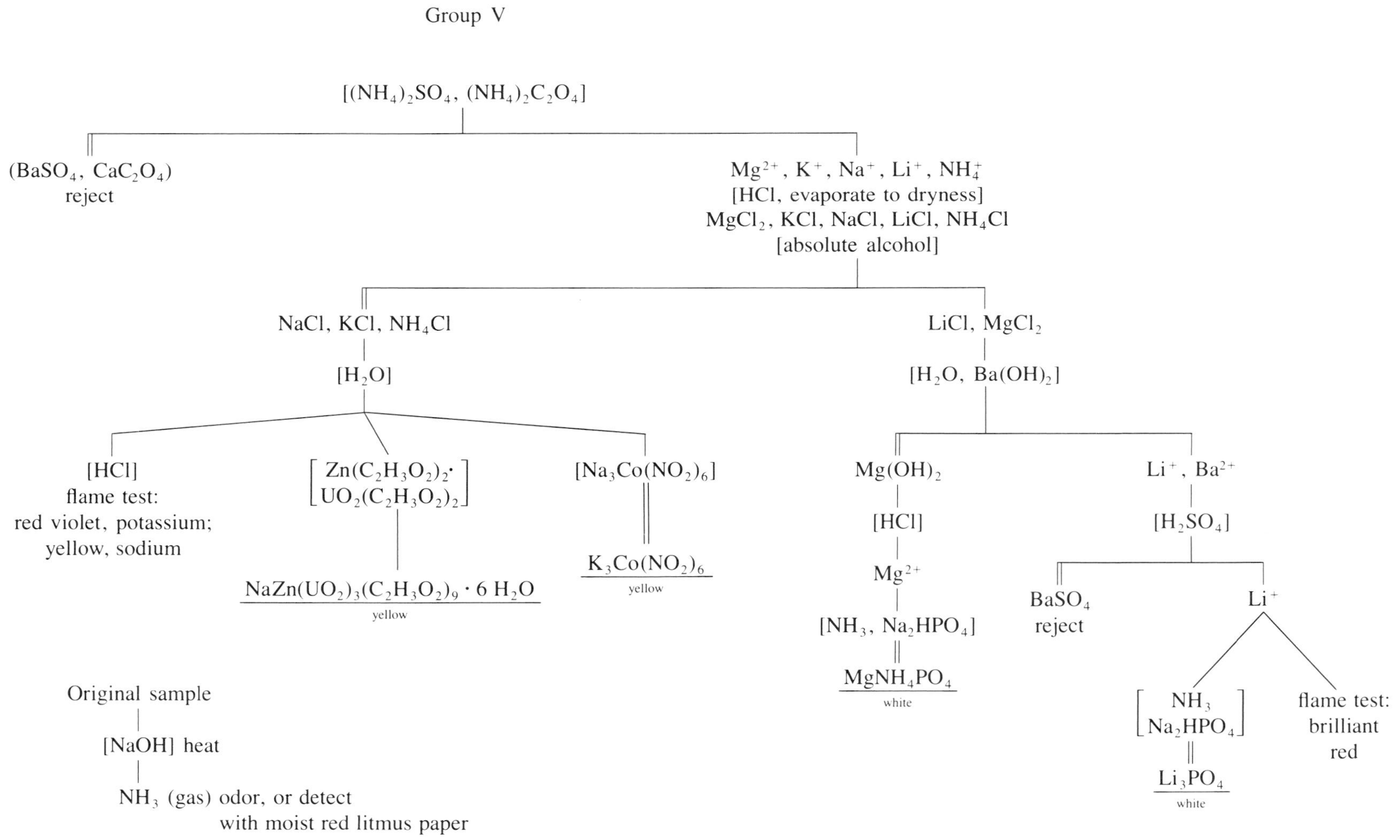

GROUP V: THE ALKALI GROUP

15

Chemistry of Separations and Tests

1. The basis for separating Mg^{2+} and Li^+ from Na^+ and K^+ is the relatively greater solubility of $MgCl_2$ and LiCl in absolute alcohol. Below are the solubilities of the chlorides in grams per 100 g absolute ethyl alcohol at 20°C:

$MgCl_2$	LiCl	NaCl	KCl
5.6	24.3	0.065	0.034

Isopropyl alcohol may be used in place of ethyl alcohol.

2. Mg^{2+} may be separated from Li^+ by precipitation of $Mg(OH)_2$ with $Ba(OH)_2$.

$$Mg^{2+} + 2\ OH^- \rightarrow \underline{Mg(OH)_2}$$

After solution of $Mg(OH)_2$ in acid, Mg^{2+} is identified by precipitation of $MgNH_4PO_4$ in ammoniacal solution with Na_2HPO_4.

$$Mg^{2+} + NH_4^+ + PO_4^{3-} \rightarrow \underset{\text{white}}{\underline{MgNH_4PO_4}}$$

3. Li^+ is identified by its brilliant red flame test, and also by precipitation of Li_3PO_4.

$$3\ Li^+ + Na_2HPO_4 + OH^- \rightarrow \underset{\text{white}}{\underline{Li_3PO_4}} + 2\ Na^+ + H_2O$$

4. Na^+ and K^+ may be identified in the presence of each other by precipitation reactions. K^+ is identified by precipitation of yellow potassium hexanitrocobaltate(III), $K_3Co(NO_2)_6$, with a fresh solution of $Na_3Co(NO_2)_6$.

$$3\ K^+ + Na_3Co(NO_2)_6 \rightarrow \underset{\text{yellow}}{\underline{K_3Co(NO_2)_6}} + 3\ Na^+$$

Na^+ may be precipitated as a finely divided yellow solid with zinc uranyl acetate reagent.

$$2\ Na^+ + 6\ Zn(C_2H_3O_2)_2 \cdot UO_2(C_2H_3O_2)_2 + 12\ H_2O \rightarrow$$
$$\underset{\text{yellow}}{\underline{2\ NaZn(UO_2)_3(C_2H_3O_2)_9 \cdot 6\ H_2O}} + 3\ Zn(C_2H_3O_2)_2 + Zn^{2+}$$

Both Na^+ and K^+ may be confirmed by flame tests. K^+ gives a reddish violet flame when viewed through a cobalt glass filter (to filter out any sodium flame). Na^+ gives a fluffy yellow flame that lasts several seconds if an appreciable amount of sodium is present in the sample. A little practice enables one to differentiate between traces of sodium and appreciable amounts of the element.

5. NH_4^+ ion is detected by liberation of NH_3 gas by treatment with a strong base. For example:

$$NH_4^+ + OH^- \rightarrow NH_3\uparrow + H_2O$$

The ammonia gas may be detected by odor or by its effect on moist red litmus.

Preliminary Tests for Group V

None. Proceed directly to the analytical procedures.

Analytical Procedures for Group V

Make up a known solution using 4 drops each of test solutions of Mg^{2+}, K^+, Li^+, Na^+, and NH_4^+. Analyze according to Procedure 13.

Procedure 13 Analysis of Group V
Solution: (From Procedure 12.) Mg^{2+}, Li^+,* K^+, Na^+, NH_4^+. Adjust the volume of solution to about 0.5 mL. If solution is not already basic, add dilute NH_3 until basic. Add 1 drop $(NH_4)_2SO_4$ and 1 drop $(NH_4)_2C_2O_4$ (1). If precipitate forms, centrifuge and discard precipitate. Treat the solution according to directions below. If Li^+ might be present, treat according to Procedure A. If Li^+ is known to be absent, treat separate portions of the solution according to Procedure B.
Procedure A: If Li^+ is to be tested for, place the solution in a small crucible or evaporating dish, add 2 drops 6 *M* HCl, and carefully evaporate to dryness. Treat residue with 2 mL absolute ethyl alcohol or isopropyl alcohol; break up the solid with a stirring rod and stir well to digest the residue. Transfer the residue and solution to a small dry test tube, then centrifuge.
Residue: NaCl, KCl, NH_4Cl. Dissolve in 2 mL H_2O. Use separate portions to test for Na^+ and K^+ according to Procedure B below.
Solution: $MgCl_2$, LiCl. Add 1 mL H_2O, then 0.5 mL $Ba(OH)_2$. Heat to near boiling. Centrifuge.
Precipitate: $Mg(OH)_2$. Dissolve with 3–4 drops dilute HCl. Add dilute NH_3 until basic; then add 2 drops Na_2HPO_4. Allow to stand for a few minutes (2). Formation of a white crystalline precipitate ($MgNH_4PO_4$) proves *magnesium*. *(continued)*

*Lithium may be omitted at the option of the instructor.

Procedure 13 *(continued)*

Solution: Li^+, Ba^{2+}. Add 3 drops dilute H_2SO_4 to precipitate Ba^{2+} as $BaSO_4$. Centrifuge and discard precipitate. Evaporate solution to about 1 mL. Use a few drops of solution for flame test. A brilliant red flame indicates Li^+. To the remainder of the solution add NH_3 until basic; add 2 drops Na_2HPO_4. A slow-forming white precipitate proves *lithium* (3).

Procedure B: If Li^+ is not present, test for Mg^{2+}, Na^+, and K^+ as follows:

Test for Mg^{2+}: To 6 drops of the solution in a test tube, add 2 drops dilute NH_3, followed by 2 drops Na_2HPO_4. Allow to stand for a few minutes (2). The formation of a white crystalline precipitate ($MgNH_4PO_4$) proves *magnesium*.

Flame Tests for Na^+ and K^+: Evaporate several drops of the solution to dryness in a crucible or evaporating dish. Moisten the residue with a drop of dilute HCl and make a flame test. Observe the flame through a cobalt glass filter. A reddish violet flame through the glass indicates *potassium*.

If the flame is a fluffy yellow (without filter) that lasts for several seconds, *sodium* is present (4).

Confirm these tests by using the *original* sample. If appreciable amounts of sodium are present, a good flame test should be obtained on the *original* sample.

Test for Na^+: To a few drops of solution in a test tube add 8 drops zinc uranyl acetate. Cool under the tap and allow to stand for 10 minutes. Centrifuge. A fine yellow precipitate confirms *sodium*.

Test for K^+: Evaporate 6 drops of the solution to dryness in an evaporating dish. Add 2–3 drops dilute HNO_3 to the residue and again evaporate to dryness. Bake the residue for several minutes with a strong flame (5).

Treat the residue with a drop of dilute $HC_2H_3O_2$ and 3–4 drops H_2O. Stir well and transfer to small test tube. Add a drop of freshly prepared sodium hexanitrocobaltate(III). Warm. A yellow precipitate forming slowly confirms *potassium*.

Test for NH_4^+: The test for ammonium ion must be made on the *original* sample, since ammonium salts have been added in the analysis. To 3 drops of the original sample (or several crystals, if the sample is solid) in the bottom of a test tube, carefully add NaOH dropwise until basic (6). Place a moistened strip of red litmus over the edge of the tube so that about one-half of it is on the inside of the tube. Heat the tube cautiously. A blue color to the litmus proves *ammonium* ion. (Instead of using the litmus above, the solution may be heated to boiling and smelled. The odor of ammonia is usually very apparent if *ammonium* ion was present.)

Notes

1. Since $BaCO_3$ and $CaCO_3$ are slightly soluble in alkaline solution, small amounts of Ca^{2+} and Ba^{2+} may appear in the filtrate containing Group V. The purpose of adding $(NH_4)_2C_2O_4$ and $(NH_4)_2SO_4$ is to remove these ions as $BaSO_4$ and CaC_2O_4. If they were not removed, they would interfere with the test for Mg by precipitation as insoluble phosphates.

2. The precipitate of $MgNH_4PO_4$ forms slowly; consequently, the solution should be allowed to stand for 5–10 minutes.

3. Li_3PO_4 is slow to form and resembles small ice crystals. Be sure solution is basic.

4. Potassium and sodium should *always* be confirmed by flame tests on the *original* sample. Evaporate a few drops of solution to near dryness for the tests.

5. Ammonium salts are volatile and hence can be removed by baking the residue of salts. If not removed, the ammonium salts interfere with the test for K^+. Freshly prepared $Na_3Co(NO_2)_6$ is imperative. Oxidation and decomposition of sodium hexanitrocobaltate(III) occur as the reagent stands. It is best to test reagent for effectiveness with a known K^+ solution before using with the unknown.

6. Carefully avoid touching the test tube with the pipet used to insert NaOH. If NaOH comes in contact with the litmus, a false test for NH_4^+ will be obtained.

Analysis of Unknown Solution over Groups IV–V

Obtain from the instructor an unknown containing ions of Groups IV–V only. Use 0.5 mL for analysis. Analyze according to Procedures 12 and 13. Record results in diagrams similar to the ones on pages 64 and 68.

Analysis of General Unknown

Obtain from your instructor an unknown over Groups I–V. Analyze according to procedures for cation analysis. Use 0.5 mL for analysis. Develop a diagram of the analysis and write equations for reactions of all ions found.

PART FOUR

ANALYSIS OF ALLOYS, SALTS, AND COMMERCIAL SUBSTANCES

ANALYSIS OF ALLOYS

16

The product obtained by melting together two or more metals and allowing the mixture to solidify is called an *alloy*. In the solid state, just as in liquids, elements vary in solubility in each other, from very soluble to very insoluble. Thus we may have in the solid form:

1. A solid solution of one or more metals, one in another, with any excess above the amount required for saturation present as crystals imbedded in the solid solution

2. Elements almost insoluble in each other in solid form, resulting in a close-packed mixture of distorted crystals

3. The formation of compounds between the metals on cooling

Common Alloys

Trade Name	Composition, %
Aluminum bronze	90 Cu, 10 Al
Babbitt	90 Sn, 7 Sb, 3 Cu
Bearing bronze	82 Cu, 16 Sn, 2 Zn
German silver	60 Cu, 25 Zn, 15 Ni
Gold coinage	90 Au, 10 Cu
Magnalium	90 Al, 10 Mg
Manganese bronze	95 Cu, 5 Mn
Nickel coinage	75 Cu, 25 Ni
Pewter	85 Sn, 6 Bi, 7 Cu, 2 Sb
Red brass	90 Cu, 10 Zn
Silver coinage	90 Ag, 10 Cu
Solder	50 Pb, 50 Sn
Type metal	82 Pb, 15 Sb, 3 Sn
Yellow brass	67 Cu, 33 Zn
Wood's metal	50 Bi, 25 Pb, 12.5 Sn, 12.5 Cd

It is evident that the composition of alloys may vary within wide limits: certain metals may be present in very small percentages in certain alloys; in others, a high percentage of these metals may be present. In the foregoing analysis of known solutions of cations, a sample was taken that contained 0.5 to 1 mg of each ion to be tested. Such an ideal case of having each of the ions present in about equal quantities is rarely found in the case of alloys. Thus one is confronted with problems of choosing the proper size of sample so that all elements will be found, and yet not so much of any one element to hinder "clean" or "complete" separation from other elements. For example, if a certain bronze alloy containing 82% Cu, 16% Sn, and 2% Zn is to be analyzed, the sample should be large enough so that zinc will be found. However, since the copper content is 41 times greater than that of zinc, too large a sample must be avoided. If a 20 mg sample of this alloy is taken at the start of the analysis, the quantity of zinc to be precipitated in Group III (0.4 mg) will be comparable with that in 1 drop of Zn^{2+} test solution. (*In general, 20 mg is a suitable size sample for an unknown alloy.*) In this same sample 16.4 mg of Cu will be precipitated in Group II (compared with 0.5 mg in a known).

Thus the results of the analysis may, and should rightly, be reported on a semiquantitative basis. An attempt should be made to classify an element as a *major ingredient*, a *minor ingredient* (a small percentage), or a *trace*. Impurities in reagents sometimes give trace tests.

Nitric acid is probably the most effective reagent in attacking alloys, since it dissolves all metals except gold and platinum to form soluble nitrates. Tin and antimony, however, form insoluble oxides.

$$Cu + 4\ HNO_3 \rightarrow Cu(NO_3)_2 + 2\ NO_2 + 2\ H_2O$$

$$Fe + 6\ HNO_3 \rightarrow Fe(NO_3)_3 + 3\ NO_2 + 3\ H_2O$$

$$Sn + 4\ HNO_3 \rightarrow \underline{SnO_2} + 4\ NO_2 + 2\ H_2O$$

$$2\ Sb + 10\ HNO_3 \rightarrow \underline{Sb_2O_5} + 10\ NO_2 + 5\ H_2O$$

Ferrous and Nonferrous Alloys

The alloys listed on page 75 may be classed as nonferrous alloys. Ferrous alloys are for the most part specialty steels, and the elements commonly alloyed with Fe are Cr, Mn, W, Mo, V, Ni, Co, Si, and C. The instructor may designate whether an unknown alloy is a ferrous or nonferrous one. Ferrous alloys are hard to dissolve in HNO_3 and should be dissolved in HCl. Consequently, if little action is noted with HNO_3, the alloy should be washed free of HNO_3 and treated with concentrated HCl. In the latter case, tin and antimony will be present in solution with other cations, and will be tested for in the regular procedures for cation analysis.

Analysis of an Unknown Alloy

Obtain an unknown alloy from your instructor and analyze it according to Procedure 14. The instructor will probably tell you whether it is a ferrous or nonferrous alloy. Record results as you did for your unknown cation solutions.

Procedure 14 Analysis of Alloy

Place about 20 mg of the alloy in the form of turnings or filings in a small beaker. Follow Procedure A for a nonferrous alloy and Procedure B for a ferrous alloy.

Procedure A: *Nonferrous.* Add 3 mL 6 *M* HNO_3 (1),(2), cover with a watch glass, and heat in a hood until brown fumes are no longer evolved. (Use more HNO_3 as necessary.) Treat by **(a)** or **(b)**.

(a) If solution is complete, place solution in an evaporating dish and boil off excess HNO_3 (use hood). The material can be taken to almost dryness—to crystal formation. Avoid further heating as nitrates decompose to water-insoluble oxides. Add 5 mL cold H_2O and 1 mL concentrated HCl. If precipitate forms, separate precipitate and analyze it for the silver group, Procedure 4.

Evaporate the filtrate (or the solution in case no precipitate was formed with HCl) almost to dryness to remove excess acid. Take up residue in 5 mL H_2O and analyze as usual for the cations of Groups II and III.

(b) If solution is incomplete, a white residue remains. (A white residue here may be hard to see against the evaporating dish background.) Add 5 mL H_2O, stir, and centrifuge. Decant the solution and later analyze it for cations according to Procedure 4.

Wash the white residue of possible SnO_2 and Sb_2O_5 at least three times with dilute HNO_3. Discard washings. Place residue in an evaporating dish; add 2 mL concentrated HCl and heat to boiling to dissolve antimony and tin oxides.

Analyze the solution possibly containing $SnCl_4$ and $SbCl_3$ as follows: Add 0.2 g oxalic acid crystals and 5 drops H_2O to 5 drops of the solution. Add H_2S (thioacetamide) to the hot solution. Reddish orange Sb_2S_3 indicates antimony. Separate the Sb_2S_3, if any, from the solution.

Boil the solution until H_2S is removed. Add dilute HCl if necessary to keep volume constant. Add a piece of Mg ribbon to the solution. Boil 2–3 minutes. Centrifuge and discard any residue. Cool the mixture and pour it into a solution of $HgCl_2$. White to grey precipitate indicates tin.

Procedure B: *Ferrous.* Add 8–10 drops 6 *M* HCl and warm until action ceases or until solution is complete. Centrifuge and discard any small residue (2). Use the solution for analysis of cation Groups II and III (3).

Notes

1. Unless the alloy decomposes H_2O, no alkali or alkaline earth metals are present. Calcium, barium, strontium, sodium, and potassium are seldom included in alloys.

2. C, P, As, and Si are sometimes found in alloys. As and P will dissolve in HNO_3, forming arsenic and phosphoric acids. C and Si will be an insoluble residue, and

should not be mistaken for Sb and Sn. Do not use too much HNO_3 or evaporation will require too much time.

3. Group I metals cannot be present since Hg_2Cl_2, $PbCl_2$, and AgCl are insoluble. The excess HCl should be removed by boiling or by neutralization with NH_3 such that *p*H may be adjusted for Group II precipitation with sulfide ion. One does not usually find Group II metals (Cu, Bi, Cd, Sb, Hg, Sn, As) in a ferrous alloy; consequently the analytical procedures for ferrous alloys will normally start with Group III.

ANALYSIS OF SALTS AND MIXTURES

17

The analysis of a salt or mixture of salts consists of the following steps:

1. Physical examination of sample
2. Solution of sample
3. Cation analysis
4. Anion analysis

Physical Examination

If an unknown is a solid, a careful observation of certain physical properties may yield valuable information about its composition. Does the substance appear homogeneous and uniform in composition? Is it crystalline or noncrystalline? Is it metallic or salt-like? Does the color indicate certain elements? These questions can be answered by careful inspection of the sample. Certain cations and anions have characteristic colors that may be indicative of the composition. Below are listed some of the more common colored ions:

Cu^{2+}	blue	CrO_4^{2-}	yellow
Ni^{2+}	green	$Cr_2O_7^{2-}$	orange-red
Co^{2+}	pink to blue	MnO_4^-	purple
Fe^{3+}	yellow (with Cl^-)	Mn^{2+}	pale pink
Cr^{3+}	green	Fe^{2+}	pale green

Selection of a Solvent

Before analyzing a solid unknown for cations and anions, it is necessary to dissolve the sample. A pinch of the sample no larger than the head of a pin should be treated with 2–3 mL of various solvents to determine which solvent to use in the preparation of a solution for cation analysis. The order of solvents is: (1) cold then hot water; (2) dilute HCl (hot if necessary); (3) dilute HNO_3 (hot if necessary); (4) hot concentrated HCl; (5) hot concentrated HNO_3; (6) hot concentrated HCl and hot concentrated HNO_3. Test the solubility of the sample in these solvents in the order listed until a suitable solvent is found. In determining the solvent to use, allow plenty of time for solution; if the solvent appears to act slowly on the sample, allow more time for solution. Certain mixtures may be partially soluble in one solvent or another; it is best to find a solvent that will completely dissolve the sample, and thus obtain only one solution for analysis. Certain substances, such as $PbSO_4$, $AgCl$, $BaSO_4$, are insoluble in the solvents listed above and require fusion with Na_2CO_3 for solution. Substances requiring special treatment for solution will not be included in this course.

After selecting a suitable solvent, dissolve a portion about the size of a small pea in as little solvent as possible and analyze for cations according to Procedure 4.

Analysis for Cations

Ordinarily the analysis for cations precedes that for anions, since, on the basis of the cation analysis, elimination of certain anions may be indicated. For example, if the anions CrO_4^{2-} or AsO_4^{3-} are present, the cations of these elements will be found in the cation analysis. Obviously, if these elements are not found in the cation analysis, the anions cannot be present, and tests for these anions need not be made. Also, if the sample is a solid, certain anions can be eliminated on the basis of solubility considerations and the cation analysis. For example, a solid that dissolves in water is found to contain Ba^{2+}. Since the barium salts of the anions SO_4^{2-}, CO_3^{2-}, SO_3^{2-}, BO_3^{3-}, AsO_4^{3-}, CrO_4^{2-}, and PO_4^{3-} are insoluble in water, these anions can be eliminated as possibilities, and anion tests need be made only for the anions that yield soluble barium salts—that is, Cl^-, Br^-, I^-, S^{2-}, NO_3^-, NO_2^-, and $C_2H_3O_2^-$.

Analysis for Anions

If cations other than the alkalies have been found in the cation analysis, they may interfere with the group tests and specific tests for the anions. Hence it will be necessary to prepare a solution free of interfering cations before making group tests and specific tests for anions. The preparation of a solution for anion analysis may be accomplished by Procedure 15. Use this prepared solution for making group tests and specific tests according to Procedures 2 and 3.

Procedure 15 Sample for Anion Analysis

Preparation of a Solution for Anion Analysis: Boil a small portion (0.2–0.3 g) of the solid unknown with 5 mL concentrated Na_2CO_3 solution. Filter, and discard the precipitate. Acidify the filtrate with HNO_3 and boil until evolution of CO_2 has ceased. Use this solution in group tests and specific tests for anions (except for carbonate, nitrate, nitrite, sulfide, and sulfite) (1).

$$\text{Cation–Anion} + Na_2CO_3 \rightarrow \underline{\text{Cation–Carbonate}} + \text{Na–Anion}$$

$$\begin{aligned}\text{Na–Anion} + HNO_3 \rightarrow NaNO_3 + H_2O &+ CO_2 \uparrow \\ &+ H_2S \uparrow \text{ or } S \\ &+ SO_2 \uparrow \\ &+ \text{H–Anion solution}\end{aligned}$$

If interfering cations are absent from the unknown, a solution for anion analysis may be prepared simply by dissolving a portion of the solid in water.

Notes

1. HNO_3 would destroy CO_3^{2-}, NO_2^-, SO_3^{2-}, and of course a solution to which HNO_3 has been added should not be tested for NO_3^-. The test for these ions is on the *original* sample.

Analysis of Unknowns

Each student should analyze at least two samples of a single salt and one sample of mixed salts. Obtain samples from the instructor and record all information as indicated on the sample data sheet that follows.

Sample Data Sheet for the Analysis of Salts and Mixtures

Number of sample ____________________

Physical properties of sample:

Color?______________________________ Crystalline or noncrystalline?__________

Homogeneous?_______________________ Soluble in H_2O?____________________

If insoluble in H_2O, in what solvent does it dissolve?__________________________

Cation analysis:

Group I ______________ Group II ______________ Group III ______________

Group IV______________ Group V______________ NH_4^+______________

Anion analysis:

H_2SO_4 treatment (what happened?)_______________________________________
On the basis of H_2SO_4 treatment and solubility considerations, can you eliminate any anions as possibilities? If so, list them:

____

Group tests for anions:

Sulfate group______________________ Halide group______________________

Specific tests for anions:

If you did not make certain specific tests, indicate why.

Sulfate___________________________ Chloride__________________________

Sulfite___________________________ Bromide__________________________

Carbonate_________________________ Iodide____________________________

Chromate____________________ Sulfide____________________

Borate____________________ Acetate____________________

Arsenate____________________ Nitrate____________________

Phosphate____________________ Nitrite____________________

Report on unknown:

ANALYSIS OF COMMERCIAL SUBSTANCES 18

Procedures have been outlined for identification of the *common metals* (singly or in admixture) under alloy analysis coupled with cation analysis.

Many *pharmaceuticals* and *industrial substances* are salts or salt mixtures, and may be tested for as such. Anion analysis identifies acids. Qualitative cation testing of solutions made from metal oxides or hydroxides identifies these types of substances.

Separation of Organic and Inorganic Matter

The scope of this text is inorganic analysis. Analysis of organic compounds is a separate branch of chemistry. At times, however, *a mixture of inorganic and organic matter* is encountered. The presence of organic matter in a sample may usually be detected by heating a small portion of the solid sample in a small test tube. Charring and burnt odor accompanies decomposition of most organic matter. By heating a larger portion of organic–inorganic mixture to a high temperature in a crucible, carbonaceous material may be volatilized to leave an inorganic residue. Consideration must be given to the fact that some inorganic matter can be vaporized, partly decomposed, or sublimed [for example, As_2O_3, $Cu(NO_3)_2$, $CaCO_3$, Hg_2Cl_2].

Detection of common inorganic substances in foods may be carried out after removal of organic matter. In testing food for possible inorganic materials that might volatilize or decompose in the ignition process, a wet oxidation of the organic matter from inorganic may be carried out. Put a 1 g sample of the dried and mascerated mixture containing organic matter in a casserole or evaporating dish, and give vigorous digestion with a hot HNO_3–H_2SO_4 mix. Finally, heat to eliminate HNO_3 to the point of obtaining SO_3 fumes. Cool the resulting solution and pour it into cold water to give a solution for inorganic analysis.

In separating some organic from inorganic matter, use is made of solvents for organic matter, such as CCl_4,* ether, turpentine, and gasoline. Subsequent to such organic removal, inorganic material can be separately analyzed.

*A safer substitute for CCl_4 is 1,1,1-trichloroethane.

Analyzing Minerals

Rocks and minerals are most commonly identified (as studied in petrography and mineralogy) from physical properties of the minerals. The elements of qualitative chemical analysis are often employed in definitely identifying minerals not readily discerned from physical properties. Since most minerals are water insoluble, the main problem in qualitative testing is dissolving them. Most minerals are oxides, mixed oxides, or salts, and acids will dissolve many of them.

The most common anions in rocks are silicates. Ground silicate minerals are first fused with excess Na_2CO_3. The fusion residue can be digested with hot water and then filtered. The residue is dissolved in acid for cation analysis.

$CaO \cdot FeO \cdot Al_2O_3 \cdot XSiO_2$ (a silicate mineral)
fuse with Na_2CO_3
digest with water

$CaCO_3$, $FeCO_3$, $Al(OH)_3$ — $Na_2O \cdot SiO_2$, Na_2CO_3

$CaCO_3$, $FeCO_3$, $Al(OH)_3$
[HCl]
solution for cation analysis

Siliceous rocks may be decomposed for analysis by treating the ground rock in a platinum crucible with HF.* SiF_4 volatilizes and the fluoride metal salt residue can be dissolved for cation analysis.

---

*HF is an especially hazardous material that must be used under a hood. Carefully avoid skin contact.

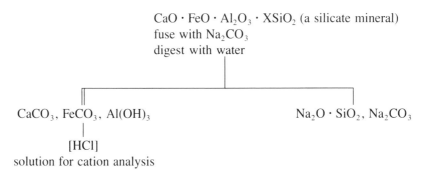

PART FIVE

PROBLEMS, EXERCISES, AND APPENDIX

PROBLEMS AND EXERCISES

Problems and exercises are arranged to conform to the usual topical sequence of a qualitative analysis course. Answers are supplied, but the student should refer to the appropriate section of the text for the basic information needed to complete the exercises.

Composition of Solutions

1. Determine the number of moles in 100 g of the following: **(a)** HCl **(b)** H_3AsO_4 **(c)** $ZnCl_2$ **(d)** $Fe_2(SO_4)_3$ **(e)** $NH_4C_2H_3O_2$ **(f)** $Cr(OH)_3$

Ans. **(a)** 2.74 **(b)** 0.70 **(c)** 0.73 **(d)** 0.25 **(e)** 1.3 **(f)** 0.97

2. Calculate the moles of the metallic element contained in the given weights of the following compounds: **(a)** 31.2 g Ag_2SO_4 **(b)** 12.27 g $BaCl_2 \cdot 2\,H_2O$

Ans. **(a)** 0.2 **(b)** 0.05

3. How many milligrams of NaOH are required to react completely with 5.6 mg of: **(a)** Fe^{2+} **(b)** Fe^{3+}

Ans. **(a)** 8.0 **(b)** 12.0

4. How many grams are contained in: **(a)** 1 mole of Hg^{2+} **(b)** 0.05 mole of Cu^{2+} **(c)** 2 moles of PO_4^{3-}

Ans. **(a)** 200.6 g **(b)** 3.177 g **(c)** 190 g

5. Indicate the weights needed for 0.1 mole quantities of each of the following: **(a)** Cd^{2+} **(b)** SO_4^{2-} **(c)** NO_3^- **(d)** Hg_2^{2+}

Ans. **(a)** 11.2 g **(b)** 9.6 g **(c)** 6.2 g **(d)** 40.1 g

6. 49.5 g of $SrBr_2$ is dissolved in 1 liter of H_2O (Sr = 87.5, Br = 80). **(a)** 49.5 g equals how many moles of $SrBr_2$? **(b)** If 100% ionized, it would form how many moles of Sr^{2+}, and how many moles of Br^-? **(c)** If the $SrBr_2$ is 100% ionized, what is the molar concentration? What is the molar concentration of Sr^{2+} and Br^-?

Ans. **(a)** 0.2 mole $SrBr_2$ **(b)** 0.2 mole Sr^{2+}, 0.4 mole Br^- **(c)** 0.2 *M* $SrBr_2$, 0.2 *M* Sr^{2+}, 0.4 *M* Br^-

7. $Al_2(SO_4)_3$ dissolves in H_2O at 20°C to the extent of 17.1 g/liter of solution. **(a)** What is the molar solubility? **(b)** What is the molar solubility of Al^{3+}? **(c)** What is the molar solubility of SO_4^{2-}?

Ans. **(a)** 0.05 *M* **(b)** 0.10 *M* **(c)** 0.15 *M*

8. 91.5 g of CdI_2 is dissolved to make 1 liter of solution. **(a)** 91.5 g equals how many moles of CdI_2 (Cd = 112.4, I = 126.9)? **(b)** If 100% ionized, it would form what molarity of Cd^{2+} and I^-?

Ans. **(a)** 0.25 mole CdI_2 **(b)** 0.25 *M* Cd^{2+}, 0.50 *M* I^-

9. Calculate the molarities of the following solutions: **(a)** 0.146 g HCl in 100 mL of solution **(b)** 2.45 g H_2SO_4 in 200 mL of solution **(c)** 49 g H_2SO_4 in 4 liters of solution **(d)** 2 g NaOH in 5 liters of solution **(e)** 3.7 g $Ca(OH)_2$ in 400 mL of solution **(f)** 32.5 g $FeCl_3$ in 500 mL of solution **(g)** 427 g $Ga_2(SO_4)_3$ in 10 liters of solution

Ans. **(a)** 0.04 *M* **(b)** 0.125 *M* **(c)** 0.125 *M* **(d)** 0.01 *M* **(e)** 0.125 *M* **(f)** 0.4 *M* **(g)** 0.1 *M*

10. Calculate the weight of solute (in grams) necessary for each of the following solutions: **(a)** 250 mL of 0.1 *M* LiOH **(b)** 2 liters of 0.05 *M* H_2SeO_4 **(c)** 2 liters of 0.05 *N* H_2SeO_4 **(d)** 100 mL of 0.1 *N* $Cr_2(SO_4)_3$ **(e)** 50 liters of 3 *N* $Cr_2(SO_4)_3$ **(f)** 5 liters of 1 *M* H_3AsO_4

Ans. **(a)** 0.6 g **(b)** 14.5 g **(c)** 7.25 g **(d)** 0.653 g **(e)** 9800 g **(f)** 710 g

11. Calculate the molarity of each of the following solutions of slightly soluble substances: **(a)** AgCN, 2.2×10^{-4} g/liter **(b)** As_2S_3, 5.2×10^{-4} g/liter

(c) $BaSO_4$, 2.4×10^{-3} g/liter **(d)** PbI_2, 0.68 g/liter

Ans. **(a)** 1.6×10^{-6} *M* **(b)** 2.1×10^{-6} *M* **(c)** 1.0×10^{-5} *M* **(d)** 1.5×10^{-3} *M*

12. Using specific gravities and percentage composition given below, calculate the approximate molarity of each of the following concentrated laboratory reagents:

	Density, g/mL	Percent by weight
(a) HCl	1.2	38
(b) HNO_3	1.42	70
(c) H_2SO_4	1.84	96
(d) NH_3	0.9	28

Ans. **(a)** 12.5 **(b)** 16 **(c)** 18 **(d)** 15

13. Calculate the volume of 0.1 *M* HCl necessary **(a)** to exactly neutralize 50 mL of 0.04 *M* KOH. **(b)** to react with 2.12 g Na_2CO_3. **(c)** to neutralize 100 mL 0.01 *M* $Ca(OH)_2$. **(d)** to react with 5 liters of 0.4 *M* $AgNO_3$.

Ans. **(a)** 20 mL **(b)** 400 mL **(c)** 20 mL **(d)** 20 liters

14. 100 mL of 0.05 *M* H_2SO_4 was neutralized with 0.1 *M* KOH, and the resulting solution was evaporated to dryness. **(a)** What volume of 0.1 *M* KOH was required? **(b)** What weight of K_2SO_4 was obtained on evaporation?

Ans. **(a)** 100 mL **(b)** 0.87 g

15. Calculate the volume of 0.2 *M* H_2SO_4 that can be prepared from 25 mL of 18 *M* H_2SO_4.

Ans. 2.25 liters

Ionization Constants of Acids and Bases

1. From the following ionization constants (K_a), indicate the strongest and weakest acids.

(a) Boric acid	7.3×10^{-10}
(b) Hypobromous acid	2.1×10^{-9}
(c) Periodic acid	2.3×10^{-2}
(d) Chromic acid	1.8×10^{-1}
(e) Thiocetic acid	4.7×10^{-4}
(f) Valeric acid	1.5×10^{-5}
(g) Glutaric acid	3.4×10^{-4}
(h) Cyanoacetic acid	3.7×10^{-3}

Ans. **(d)** strongest and **(a)** weakest

2. Calculate the H^+ ion concentration of 0.1 *M* formic acid ($HCHO_2$) solution (K_a for formic acid = 1.8×10^{-4}).

Ans. 4.2×10^{-3} *M*

3. Calculate the value of K_a for HNO_2 if a 0.1 *M* solution is 7% ionized.

Ans. 5×10^{-4}

4. A 0.5 *M* solution of the base MOH is found to be 1×10^{-4}% ionized. Calculate K_b.

Ans. 5×10^{-13}

5. A 1.0 *M* solution of a certain base, MOH, is 15% ionized into M^+ and OH^-. Calculate K_b.

Ans. 2.6×10^{-2}

6. The base ionization constant for hydrazine (N_2H_4) is 9.8×10^{-7}. What is the OH^- concentration in 1.0 *M* solution?

$$N_2H_4 + H_2O \rightleftharpoons N_2H_5^+ + OH^-$$

Ans. 1×10^{-3} *M*

7. A weak acid, HA, is 2.0% ionized in 0.2 *M* aqueous solution. Calculate K_a.

Ans. 8.2×10^{-5}

8. K_a for HPO_3 = 5.0×10^{-5}. Calculate the concentration of H^+ in 0.005 *M* solution.

Ans. 5×10^{-4} *M*

9. Calculate the OH^- ion concentration in 0.1 *M* NH_3 solution (K_b for NH_3 = 1.8×10^{-5}).

$$NH_3 + H_2O \rightleftharpoons NH_4^+ + OH^-$$

Ans. 1.3×10^{-3} *M*

10. Underscore the most abundant species (omit H_2O) and encircle the least abundant species in an aqueous solution made by dissolving 1 mole of $NaC_2H_3O_2$ in 1 liter of 0.1 *M* $HC_2H_3O_2$.

H_3O^+ or H^+, $C_2H_3O_2^-$, $HC_2H_3O_2$, Na^+, OH^-

Ans. $C_2H_3O_2^-$ most abundant, OH^- least abundant

11. Calculate the approximate H^+ ion concentration in 1 liter of 0.2 *M* acetic acid to which 0.2 mole of sodium acetate has been added (K_a for $HC_2H_3O_2$ = 1.8×10^{-5}).

Ans. 1.8×10^{-5} *M*

12. Explain how a buffer solution of NH_4Cl and NH_3 would act toward the addition of a strong **(a)** acid and **(b)** base.

13. Calculate the OH^- ion concentration in 1 liter of 0.25 *M* NH_3 that contains 0.2 moles NH_4Cl (K_b for NH_4OH = 1.8×10^{-5}).

Ans. 2.25×10^{-5} *M*

14. Calculate the approximate OH^- ion concentration in a solution made by adding 29 g of NH_4I to 1 liter of 2 *M* NH_3 (K_b for NH_4OH = 1.8×10^{-5}).

Ans. 1.8×10^{-4} *M*

15. A hypothetical weak base ionizes according to the equation

$$MOH \rightleftharpoons M^+ + OH^-$$

A salt (M_2SO_4) that is 100% ionized is added. The concentration of the base is 0.1 mole/liter, and the concentration of the M_2SO_4 is 0.05 mole/liter (K_b for the weak base = 1×10^{-6}). Calculate **(a)** the OH^- ion concentration and **(b)** the H^+ ion concentration.

Ans. **(a)** 1×10^{-6} *M* **(b)** 1×10^{-8} *M*

16. Given K_a for $HC_2H_3O_2 = 1.8 \times 10^{-5}$, **(a)** calculate K_b for potassium acetate, and **(b)** calculate the approximate OH^- ion concentration in a 0.1 *M* solution of $KC_2H_3O_2$.

Ans. **(a)** 5×10^{-10} **(b)** 7×10^{-6}

*p*H and *p*OH

1. A solution contains 5×10^{-6} moles H^+ ion/liter. **(a)** Is the solution acid or basic? **(b)** What is the *p*H of the solution?

Ans. **(a)** acid **(b)** 5.3

2. **(a)** What is the *p*H of a solution in which the H^+ ion concentration is 4×10^{-4} *M*? **(b)** What is the *p*OH of this solution?

Ans. **(a)** 3.4 **(b)** 10.6

3. Calculate the concentration of H^+ and OH^- in 0.02 *M* NH_3 ($K_b = 1.8 \times 10^{-5}$).

Ans. $[OH^-] = 6 \times 10^{-4}$ *M*; $[H^+] = 1.7 \times 10^{-11}$ *M*

4. Considering hydrolysis, indicate whether water solutions of the following salts would act acidic, basic, or neutral. **(a)** NaCN **(b)** Na_3PO_4 **(c)** NH_4Cl **(d)** $Mg(C_2H_3O_2)_2$ **(e)** $ZnCl_2$ **(f)** $Ca(NO_3)_2$ **(g)** $CdCl_2$ **(h)** $MnSO_4$ **(i)** $CaBr_2$

Ans. Acidic: **c**, **e**, **g**, and **h** slightly acid. Basic: **a**, **b**, **d**. Neutral: **f**, **i**.

5. The H^+ concentration of an acetic acid solution is 0.00030 *M*. What is the *p*H of the solution?

Ans. 3.5

6. Calculate the *p*H of 0.01 *M* HBrO ($K_a = 2.1 \times 10^{-9}$).

Ans. 5.8

7. A 0.02 *M* solution of a weak organic acid (HA) has a *p*H of 5.9. Calculate K_a for the weak acid.

Ans. 7.5×10^{-11}

8. What is the hydrogen ion concentration of a solution that has a *p*H of 9.77?

Ans. 1.7×10^{-10} *M*

9. Assuming 100% ionization, *estimate* the *p*H of the following aqueous solutions. **(a)** 0.1 *M* HCl **(b)** 0.01 *M* NaOH **(c)** 0.5 *M* NaOH **(d)** 0.02 *M* HCl

Ans. **(a)** 1 **(b)** 12 **(c)** 13.7 **(d)** 1.7

10. Calculate the *p*H of 1.0 *M* benzoic acid (HBe) ($K_a = 6.0 \times 10^{-5}$).

Ans. 2.1

11. To buffer the solution in the previous exercise at a *p*H of 4.0, what weight of sodium benzoate (molecular weight = 144) should be added to 1 liter of 1.0 *M* benzoic acid?

Ans. 92 g

12. What will be the *p*OH and *p*H of a buffer solution made up of 1.0 *M* N_2H_4 and 0.50 *M* N_2H_5Cl [K_b (N_2H_4) = 9.8×10^{-7}]?

Ans. *p*OH = 5.7; *p*H = 8.3

13. What is the approximate *p*H of a buffer solution made by mixing 1 liter of 0.3 *M* $HC_2H_3O_2$ and 1 liter of 0.6 *M* $NaC_2H_3O_2$ ($K_a = 1.8 \times 10^{-5}$)?

Ans. 5.05

14. Enough NH_4Cl is added to a 0.1 *M* solution of NH_3 to make the solution 0.2 *M* with respect to NH_4^+. What is the OH^- ion concentration, *p*OH, and *p*H of the solution ($K_b = 1.8 \times 10^{-5}$).

Ans. *p*H = 8.95

15. K_a for the acid HA is 4.0×10^{-6}. Calculate the H^+ ion concentration and the *p*H of a 1 *M* solution.

Ans. 2×10^{-3} *M*; 2.7

16. A solution of NH_3 was found to have an OH^- ion concentration of 2.0×10^{-4} *M*. What is the *p*H of this solution?

Ans. 10.3

17. What is the *p*H of a solution that contains: **(a)** 0.046 *M* H^+ **(b)** 0.0002 *M* H^+ **(c)** 6.2×10^{-8} *M* H^+

Ans. **(a)** 1.34 **(b)** 3.7 **(c)** 7.2

18. Indicate whether aqueous solutions of the following compounds are acidic, basic, or neutral. **(a)** NH_4Cl **(b)** $NaC_2H_3O_2$ **(c)** Na_2SO_4 **(d)** K_2CO_3 **(e)** $BaCl_2$ **(f)** NH_4NO_3 **(g)** $ZnSO_4$ **(h)** NaCN **(i)** $Cu(NO_3)_2$ **(j)** K_3PO_4 **(k)** LiCl

Ans. Acidic: **a**, **f**, **g**, and **i** slightly. Basic: **b**, **d**, **h**, **j**. Neutral: **c**, **e**, **k**.

19. Calculate the *p*OH and *p*H of the following solutions. **(a)** 0.0063 *M* OH^- **(b)** 0.132 *M* OH^-

Ans. **(a)** *p*OH = 2.2; *p*H = 11.8 **(b)** *p*OH = 0.9; *p*H = 13.1

20. $Ca(OH)_2$ is dissolved in H_2O to the extent of 0.0092 g/liter. This solution, sometimes called limewater, can be assumed to be 100% ionized. What is the **(a)** OH^- molar concentration, and **(b)** the *p*OH of the solution?

Ans. **(a)** 2.5×10^{-4} *M* **(b)** 3.6

Solubility Products

1. The solubility products (K_{SP}) of several compounds are given below.

Compound	K_{SP}
(a) $BaCrO_4$	8.5×10^{-11}
(b) FeS	4×10^{-19}
(c) CdS	1×10^{-28}
(d) HgS	1.6×10^{-54}
(e) BaC_2O_4	1.5×10^{-8}
(f) $CaCO_3$	4.7×10^{-9}
(g) $AgBr$	5×10^{-13}
(h) NiS	3×10^{-21}

List the compounds according to increasing solubility, placing the least soluble compound first.

Ans. **(d)**, **(c)**, **(h)**, **(b)**, **(g)**, **(a)**, **(f)**, **(e)**

2. Calculate the K_{SP} of the following salts.

Salt	Solubility in moles/liter
(a) $Ca(OH)_2$	0.021
(b) $AgCN$	1.6×10^{-6}
(c) Ag_2CO_3	1.2×10^{-4}
(d) $AgIO_3$	3.1×10^{-4}
(e) PbF_2	2.6×10^{-3}
(f) Ag_3AsO_4	2.4×10^{-5}
(g) Ag_2SO_4	2.5×10^{-2}
(h) CaF_2	1.7×10^{-4}
(i) $CaCO_3$	0.00013

Ans. **(a)** 3.6×10^{-5} **(b)** 2.6×10^{-12} **(c)** 6.9×10^{-12} **(d)** 9.6×10^{-8} **(e)** 7×10^{-8} **(f)** 9×10^{-18} **(g)** 6.3×10^{-5} **(h)** 2×10^{-11} **(i)** 1.7×10^{-8}

3. From the solubilities of the following salts, calculate the K_{SP}.

Salt	Solubility in g/liter
(a) $PbCO_3$	1.1×10^{-3}
(b) $SrSO_4$	0.114
(c) $CaSO_4$	1.93
(d) Ag_2CrO_4	3.2×10^{-2}

Ans. **(a)** 1.6×10^{-11} **(b)** 3.8×10^{-7} **(c)** 2×10^{-4} **(d)** 3.5×10^{-12}

4. Calculate the K_{SP} of the following salts.

Salt	Solubility in mg/mL
(a) $AgCN$	2.1×10^{-4}
(b) CaC_2O_4	6.8×10^{-3}
(c) As_2S_3	5.2×10^{-4}
(d) $PbCrO_4$	7×10^{-4}

Ans. **(a)** 2.6×10^{-12} **(b)** 2.8×10^{-9} **(c)** 4.4×10^{-27} **(d)** 4×10^{-12}

5. The solubility of Ag_3AsO_4 is 8.5×10^{-3} g/liter. Calculate the K_{SP}.

Ans. 2.8×10^{-18}

6. The solubility of the salt MX_2 is 3.0×10^{-4} moles/liter. Calculate the K_{SP}.

Ans. 1.1×10^{-10}

7. The solubility of $Mg(OH)_2$ is approximately 0.009 g/liter. Calculate the K_{SP}.

Ans. 1.5×10^{-11}

8. The K_{SP} for $BaSO_4$ is 1.5×10^{-9}. **(a)** What is the molar concentration of Ba^{2+} in a saturated solution? **(b)** What is the number of grams of Ba^{2+} per liter of saturated solution? **(c)** What is the solubility of $BaSO_4$ in grams per liter?

Ans. **(a)** 3.9×10^{-5} M **(b)** 5.3×10^{-3} **(c)** 9.2×10^{-3}

9. From the following table of K_{SP} values determine the approximate solubility of the compound in moles per liter.

Compound	K_{SP}
(a) $MgCO_3$	1×10^{-5}
(b) MnS	7×10^{-16}
(c) $Ni(OH)_2$	1.6×10^{-16}
(d) PbI_2	8.3×10^{-9}
(e) Ag_2SO_4	1.2×10^{-5}

Ans. **(a)** 0.003 **(b)** 2.6×10^{-8} **(c)** 3.4×10^{-6} **(d)** 1.3×10^{-3} **(e)** 0.014

10. K_{SP} for Ag_2CO_3 is 8.2×10^{-12}. If a solution containing 4×10^{-5} M Ag^+ had 4×10^{-5} M CO_3^{2-} added to it, would precipitation occur? Show calculations to explain your answer.

Ans. No

11. K_{SP} for $CaSO_4 = 2.4 \times 10^{-5}$. If 0.05 mole of Na_2SO_4 is added to 1 liter of solution containing 4×10^{-5} mole of calcium chloride, will $CaSO_4$ precipitate? Give a reason for your answer.

Ans. No

12. Calculate the minimum concentration of Cl^- necessary to precipitate AgCl from a solution in which the concentration of Ag^+ is 2.4×10^{-4} M (K_{SP} for AgCl $= 1.7 \times 10^{-10}$).

Ans. 7×10^{-7} M

13. K_{SP} for $Fe(OH)_3 = 6 \times 10^{-38}$. If a solution has a Fe^{3+} concentration of 1×10^{-3} M, what maximum concentration of OH^- can be present without precipitating $Fe(OH)_3$?

Ans. 3.9×10^{-12}

14. Na_2SO_4 was added to 0.01 M $BaCl_2$ until the concentration of SO_4^{2-} was 0.0050 M. What concentration

of Ba^{2+} remained in solution? K_{SP} for $BaSO_4 = 1.5 \times 10^{-9}$.

Ans. 3×10^{-7} *M*

15. An aqueous solution is 0.005 *M* $AgNO_3$. Solid NaCl is added until Na^+ in solution is 1.0 *M*. What is the concentration of Ag^+ left in solution? K_{SP} for AgCl $= 1.7 \times 10^{-10}$.

Ans. 1.7×10^{-10} *M*

16. Equal volumes of 3.0×10^{-5} *M* $Ba(NO_3)_2$ and 2.0×10^{-5} *M* Na_2SO_4 are mixed. Will precipitation occur? Explain. K_{SP} for $BaSO_4 = 1.5 \times 10^{-9}$.

Ans. No

17. K_{SP} for $Ag_3PO_4 = 1.8 \times 10^{-18}$. Calculate the minimum concentration of PO_4^{3-} necessary to start precipitation of Ag_3PO_4 from a 0.05 *M* solution of $AgNO_3$.

Ans. 1.4×10^{-14} *M*

18. What would be the minimum molar concentration of S^{2-} that would have to be present to start precipitation of Ag_2S from a 0.05 *M* solution of $AgNO_3$? K_{SP} for $Ag_2S = 5.5 \times 10^{-51}$.

Ans. 2.2×10^{-48} *M*

19. A solution in equilibrium with a precipitate of $BaSO_4$ was found to contain 3.0×10^{-4} moles Ba^{2+} per liter and 5.7×10^{-6} moles SO_4^{2-} per liter of solution. Calculate the K_{SP} for $BaSO_4$.

Ans. 1.7×10^{-9}

20. K_{SP} for CdS is 1.0×10^{-28}. A solution of $CdCl_2$ was found to contain 1.2×10^{-4} *M* Cd^{2+}. What concentration of S^{2-} must be added to start precipitation?

Ans. 8×10^{-25} *M*

21. A saturated solution of $SrSO_4$ contains 0.081 g $SrSO_4$/500 mL of solution. Calculate the K_{SP} for $SrSO_4$.

Ans. 7.6×10^{-7}

22. Would $Al(OH)_3$ precipitate if the Al^{3+} ion concentration were 2.0×10^{-12} mole/liter and OH^- ion were added to make a OH^- ion concentration of 1.0×10^{-5} *M*? Explain your answer with calculations ($K_{SP} = 5 \times 10^{-33}$).

Ans. Yes

23. K_{SP} for $Ag_2CrO_4 = 1.9 \times 10^{-12}$. If the CrO_4^{2-} concentration in a solution is 1.3×10^{-4} *M*, what is the molar concentration of Ag^+ in contact with solid Ag_2CrO_4?

Ans. 1.2×10^{-4} *M*

24. K_{SP} for $BaCrO_4 = 8.5 \times 10^{-11}$. **(a)** What is the Ba^{2+} molarity in a $BaCrO_4$ saturated solution? **(b)** What is the actual weight in grams of Ba^{2+} in a liter of saturated $BaCrO_4$ solution?

Ans. **(a)** 9.2×10^{-6} *M* **(b)** 1.3×10^{-3} g/L

25. Calculate the K_{SP} for $Th(SeO_4)_2$, given that 0.52 g dissolve per 100 mL.

Ans. 4×10^{-6}

26. The salt thallium (I) chloride (TlCl) is soluble in H_2O at 25°C to the extent of 3.60 g/liter. Calculate the K_{SP}.

Ans. 2.3×10^{-4}

27. Given a 0.04 *M* solution of $AgNO_3$. K_{SP} of $Ag_3AsO_4 = 4 \times 10^{-18}$. What minimum concentration of AsO_4^{3-} would have to be attained to just start precipitation of Ag_3AsO_4?

Ans. 6×10^{-14} *M*

28. Calculate the approximate solubility (in moles per liter) of the compounds listed in the solutions indicated. Obtain the solubility products needed from Appendix F.2.

Compound	Solution
(a) AgBr	2×10^{-5} *M* NaBr
(b) $BaSO_4$	0.1 *M* Na_2SO_4
(c) AgCl	0.1 *M* HCl

Ans. **(a)** 2.5×10^{-8} *M* **(b)** 1.5×10^{-8} *M* **(c)** 1.7×10^{-9} *M*

29. On saturation of 0.3 *M* HCl with H_2S, the S^{2-} concentration is approximately 1.2×10^{-22} moles/liter. See Appendix F.2 for K_{SP} values. **(a)** What concentration of each of the ions Cu^{2+}, Cd^{2+}, Mn^{2+}, Fe^{2+} will be needed to just start precipitation of the sulfide in each instance? **(b)** Based on the values in **(a)**, explain the separation of Cu^{2+} and Cd^{2+} in Group II from Mn^{2+} and Fe^{2+} in Group III in the cation analysis.

Ans. **(a)** 6.7×10^{-15} *M* Cu^{2+}, 3.3×10^{3} *M* Fe^{2+}

30. (a) If the S^{2-} ion concentration of a solution that is 0.3 *M* in HCl, and saturated with H_2S, is 1.2×10^{-22}, at what concentration of Pb^{2+} will PbS just begin to precipitate? See Appendix F.2 for K_{SP} values. **(b)** At what concentration of Zn^{2+} will ZnS just begin to precipitate?

Ans. **(a)** 5.8×10^{-7} *M* **(b)** 2.1 *M*

31. Can a solution of $Cd(NO_3)_2$ in H_2O that will yield a precipitate with H_2S be made up (as in Problem 29)? Show calculations to verify your answer.

Ans. Yes

32. Can a solution of $Mn(NO_3)_2$ in H_2O that will yield a precipitate with H_2S be made up (as in Problem 29)? Show calculations to verify your answer.

Ans. No

33. The *p*H of a saturated solution of $Mg(OH)_2$ is 10.5. Calculate an approximate value of K_{SP} from this.

Ans. 1.6×10^{-11}

34. K_{SP} for $Mg(OH)_2 = 8.9 \times 10^{-12}$ and K_{SP} for $BaCO_3 = 1.6 \times 10^{-9}$. How does the molar solubility of $Mg(OH)_2$ compare with that of $BaCO_3$?

Ans. $\frac{1.6 \times 10^{-4}}{4 \times 10^{-5}}$ or 4.0 to 1.0

Equations

1. Balance the following equations by considering electron transfer or change in oxidation numbers. **(a)** $Ag_2S + HNO_3 \rightarrow Ag_2SO_4 + NO_2 + H_2O$ **(b)** $Ce_2(SO_4)_3 + H_2SO_4 + KMnO_4 \rightarrow Ce(SO_4)_2 + K_2SO_4 + MnSO_4 + H_2O$ **(c)** $HBr + NaMnO_4 + H_2SO_4 \rightarrow MnSO_4 + NaHSO_4 + Br_2 + H_2O$ **(d)** $H_2S + HBr + K_2Cr_2O_7 \rightarrow S + KBr + CrBr_3 + H_2O$ **(e)** $Mn(NO_3)_2 + HNO_3 + Pb_3O_4 \rightarrow Pb(NO_3)_2 + HMnO_4 + H_2O$ **(f)** $KI + KIO_3 + HC_2H_3O_2 \rightarrow KC_2H_3O_2 + I_2 + H_2O$ **(g)** $SbH_3 + HMnO_4 + H_2SO_4 \rightarrow MnSO_4 + H_3SbO_4 + H_2O$ **(h)** $V(SO_4)_2 + KMnO_4 + H_2O \rightarrow KHSO_4 + MnSO_4 + (VO_2)SO_4 + H_2SO_4$ **(i)** $Mo_2O_3 + KMnO_4 + H_2SO_4 \rightarrow MoO_3 + K_2SO_4 + MnSO_4 + H_2O$ **(j)** $SnCl_2 + H_2SO_3 + HCl \rightarrow SnCl_4 + H_2S + H_2O$

2. Balance the following equations, noting that more than two elements change in oxidation number. **(a)** $As_2S_3 + HNO_3 \rightarrow As_2O_5 + NO + H_2SO_4 + H_2O$ **(b)** $Zn + Ag_3AsO_4 + H_2SO_4 \rightarrow ZnSO_4 + Ag + AsH_3 + H_2O$ **(c)** $MnI_2 + Pb_3O_4 + HNO_3 \rightarrow HMnO_4 + Pb(IO_3)_2 + Pb(NO_3)_2$ **(d)** $As_2S_3 + Mn(NO_3)_2 + Na_2CO_3 \rightarrow Na_3AsO_4 + Na_2MnO_4 + Na_2SO_4 + CO_2 + NO$ **(e)** $Sb_2S_3 + Na_2CO_3 + NaNO_3 \rightarrow Na_3SbO_4 + Na_2SO_4 + CO_2 + NO$

3. Balance the following ionic equations by considering change in oxidation number or electron transfer. **(a)** $Sn^{2+} + Fe^{3+} \rightarrow Sn^{4+} + Fe^{2+}$ **(b)** $Cu + H^+ + NO_3^- \rightarrow Cu^{2+} + NO + H_2O$ **(c)** $H_2S + MnO_4^- + H^+ \rightarrow SO_4^{2-} + Mn^{2+} + H_2O$ **(d)** $C_2O_4^{2-} + Cr_2O_7^{2-} + H^+ \rightarrow Cr^{3+} + CO_2 + H_2O$ **(e)** $As_4 + H^+ + NO_3^- \rightarrow AsO_3^- + NO + H_2O$ **(f)** $Bi(OH)_3 + SnO_2^{2-} \rightarrow Bi + SnO_3^{2-} + H_2O$ **(g)** $Fe + H^+ + NO_3^- \rightarrow Fe^{3+} + NO_2 + H_2O$ **(h)** $Zn + H^+ + NO_3^- \rightarrow Zn^{2+} + NH_4^+ + H_2O$ **(i)** $Ce^{4+} + AsO_3^{3-} + H^+ \rightarrow AsO_3^- + Ce^{3+} + H_2O$ **(j)** $Pt + H^+ + NO_3^- + Cl^- \rightarrow PtCl_6^{2-} + NOCl + H_2O$

4. Balance the following by the ion-electron (half-reaction) method. Balance each half-reaction atomically and electrically; balance the gain and loss of electrons, then add the two half-reactions to obtain the balanced overall equation. Make all cancellations possible.

(a) $Sn^{2+} \rightarrow Sn^{4+}$
$Ce^{4+} \rightarrow Ce^{3+}$

(b) $CdS \rightarrow Cd^{2+} + S$
$H^+ + NO_3^- \rightarrow NO + H_2O$

(c) $Bi_2S_3 \rightarrow Bi^{3+} + S$
$H^+ + NO_3^- \rightarrow NO + H_2O$

(d) $As_2S_3 + S^{2+} \rightarrow AsS_4^{3-}$
$S_2^{2-} \rightarrow S^{2-}$

(e) $C_2O_4^{2-} \rightarrow CO_2$
$H^+ + MnO_4^- \rightarrow H_2O + Mn^{2+}$

(f) $S_2O_3^{2-} + H_2O \rightarrow SO_4^{2-} + H^+$
$Br_2 \rightarrow Br^-$

5. Complete and balance the following equations by the ion-electron (half-reaction) method. Where necessary, add H^+, OH^-, or H_2O to complete and balance a half-reaction. Do not use H^+ and OH^- together. Make all cancellations possible.

(a) $AsO_2^- \rightarrow AsO_4^{3-}$
$H^+ + MnO_4^- \rightarrow H_2O + Mn^{2+}$

(b) $Bi(OH)_3 \rightarrow Bi + OH^-$
$SnO_2^{2-} \rightarrow SnO_3^{2-}$

(c) $Cr^{3+} \rightarrow Cr_2O_7^{2-} + H^+$
$ClO_3^- \rightarrow ClO_2$

(d) $Zn \rightarrow Zn(OH)_4^{2-}$
$NO_3^- \rightarrow NH_3$

(e) $C_2H_5OH \rightarrow CH_3CHO$
$Cr_2O_7^{2-} + H^+ \rightarrow Cr^{3+}$

6. Balance the following equations (any method). **(a)** $V^{2+} + MnO_4^- + H^+ \rightarrow Mn^{2+} + VO_2^{2+} + H_2O$ **(b)** $NO_3^- + Al + OH^- + H_2O \rightarrow NH_3 + AlO_2^-$ **(c)** $PbSO_4 + Na_2CO_3 + C \rightarrow Pb + Na_2SO_4 + CO_2$ **(d)** $As_2S_5 + ClO^- \rightarrow AsO_4^{3-} + SO_4^{2-} + Cl^-$ **(e)** $Sb_2O_3 + Zn + H^+ \rightarrow SbH_3 + Zn^{2+} + H_2O$ **(f)** $CuNCS + MnO_4^- + H^+ \rightarrow Mn^{2+} + CN^- + Cu^{2+} + S + H_2O$ **(g)** $NpO_2^+ + H^+ \rightarrow Np^{4+} + NpO_2^{2+} + H_2O$ **(h)** $VO^+ + H^+ + Cr_2O_7^{2-} \rightarrow VO_3^- + Cr^{3+} + H_2O$ **(i)** $Sn^{2+} + ClO_4^- + H^+ \rightarrow Sn^{4+} + Cl^- + H_2O$ **(j)** $Pd + H^+ + Cl^- + NO_3^- \rightarrow PdCl_4^{2-} + NOCl + H_2O$

7. Complete and balance. **(a)** $AgCl + NH_3 \rightarrow$ **(b)** $Bi^{3+} + H_2S \rightarrow$ **(c)** $K^+ + ClO_4^- \rightarrow$ **(d)** $CO_3^{2-} + H^+$ (excess) $\rightarrow$ **(e)** $Ba^{2+} + SO_3^{2-} \rightarrow$ **(f)** $Al^{3+} + OH^-$ (excess) $\rightarrow$ **(g)** $PbS + HNO_3$ (dilute) $\rightarrow$ **(h)** $Cd^{2+} + NH_3 \rightarrow$ **(i)** $SnS_3^{2-} + H^+ \rightarrow$ **(j)** $Sn^{4+} + Mg \rightarrow$

8. From the table of standard electrode potentials in Appendix E determine whether or not the following reactions will occur. **(a)** $Mn + Sn^{2+} \rightarrow Mn^{2+} + Sn$ **(b)** $2\,Ag + 4\,H^+ + SO_4^{2-} \rightarrow 2\,Ag^+ + H_2SO_3 + H_2O$ **(c)** $2\,Na^+ + 2\,F^- \rightarrow F_2 + 2\,Na$ **(d)** $6\,I^- + Cr_2O_7^{2-} + 14\,H^+ \rightarrow 3\,I_2 + 2\,Cr^{3+} + 7\,H_2O$

Laboratory Techniques

1. What is meant by a "blank" test?

2. A precipitate is to be washed to reduce the amount of soluble reagent surrounding it. Consider the following possibilities: **(a)** one washing with a 3 mL portion of H_2O **(b)** three washings with 1 mL portions of H_2O Assume 0.5 mg of the reagent is present at the beginning of the washings. If this reagent is soluble to the extent of 0.5 mg/mL, and if after each washing 0.1 mL of liquid remains in the tube with the precipitate, calculate the milligrams of the reagent left after each washing.

Ans. **(a)** 0.017 mg; **(b)** after first washing 0.05 mg, after second washing 0.0045 mg, after third washing 0.00041 mg

3. In Problem 2, how many more times efficient is the triple washing with a total of 3 mL of wash liquid than was the single washing with 3 mL?

Ans. about 42

Analysis for Anions

1. Explain clearly the necessity of making a prepared solution for anion analysis.

2. A prepared solution for anion analysis gave no precipitate with $BaCl_2$ + NH_3 or with $AgNO_3$ + HNO_3. Which of the following anions cannot be eliminated from specific tests? **(a)** PO_4^{3-} **(b)** Br^- **(c)** AsO_4^{3-} **(d)** NO_3^- **(e)** SO_4^{2-}

3. An unknown solution containing only 1 anion gives a white precipitate with $BaCl_2$ and NH_3. The unknown solution also liberates a gas when treated with dilute HCl. When bubbled into $Ca(OH)_2$, the gas gives a white precipitate. A portion of the unknown when treated with $BaCl_2$ and HCl gives no precipitate, but, when H_2O_2 is added to this solution, a white precipitate forms. What is the anion in the unknown?

4. In the chart below, indicate what happens to each of the anions when its sodium salt is **(a)** treated with concentrated H_2SO_4. **(b)** dissolved in H_2O, and NH_3 and $BaCl_2$ are added. **(c)** dissolved in H_2O, and HNO_3 and $AgNO_3$ are added.

(If no reactions occur, indicate.)

Anion	H_2SO_4 Treatment	$BaCl_2$ + NH_3	$AgNO_3$ + HNO_3
Br^-			
NO_3^-			
SO_4^{2-}			
Cl^-			
PO_4^{3-}			
CO_3^{2-}			
I^-			
$C_2H_3O_2^-$			
AsO_4^{3-}			
S^{2-}			

Cation Analysis

Group I

1. In the precipitation of Group I with HCl, what would happen if the solution were very dilute?

2. If you were making up a solution of cations of the silver group, which salts would you use?

3. If the precipitate of Group I was insoluble in hot H_2O but completely soluble in NH_3, what conclusions could be drawn?

4. Using a single reagent, how could you distinguish between the pairs listed below? Indicate the reaction for each. **(a)** AgCl and $ZnCl_2$ (solids) **(b)** $PbCl_2$ and Hg_2Cl_2 (solids) **(c)** $Cd(NO_3)_2$ and $AgNO_3$ (solutions) **(d)** HCl and HNO_3 (solutions) **(e)** $Hg_2(NO_3)_2$ and $Hg(NO_3)_2$ (solutions)

Group II

1. In the precipitation of Group II, what would happen if the HCl were too concentrated? Too dilute?

2. Why is the precipitation carried out in 0.3 *M* HCl?

3. How is complete precipitation determined?

4. What conclusions may be drawn if the precipitate of Group II sulfides is completely soluble in Na_2S_2?

5. Why use dilute HNO_3 rather than concentrated HNO_3 in dissolving the sulfides of the copper group?

6. In precipitating $PbSO_4$ in Group II, why is it necessary to remove HNO_3 from the solution by evaporating to fumes of SO_3?

7. What is the basis for the separation of the following ions? **(a)** lead from copper, cadmium, and bismuth **(b)** bismuth from copper and cadmium **(c)** copper from cadmium

8. Give the formula for a reagent that will **(a)** dissolve CuS but not HgS. **(b)** dissolve $Cu(OH)_2$ but not $Bi(OH)_3$. **(c)** form a precipitate with Bi^{3+} but not with Cd^{2+}. **(d)** dissolve $PbSO_4$. **(e)** form a precipitate with Cu^{2+} but not with Cd^{2+}.

9. What precautions should be taken in evaporating concentrated H_2SO_4 to fumes of SO_3?

10. What conclusions may be drawn if **(a)** only a white, slowly settling precipitate is produced on acidifying the Na_2S_2 solution? **(b)** the precipitate of this group is completely soluble in concentrated HCl? **(c)** the precipitate of this group is yellow in color?

11. Give the formula for a reagent that will **(a)** dissolve As_2S_5 but not HgS. **(b)** dissolve Sb_2S_5 but not As_2S_5. **(c)** form a precipitate with $HgCl_2$ but not with $SnCl_2$.

12. Diagram a scheme for the separation and confirmation of the following mixture of ions in H_2O. (Other cations may be assumed to be absent.) **(a)** Hg_2^{2+} **(b)** Cu^{2+} **(c)** Sb^{3+}

Group III

1. What would happen if NaOH rather than NH_3 were added in the precipitation of Group III with H_2S?

2. How should excess Na_2O_2 be disposed of?

3. What conclusions can be drawn if **(a)** the precipitate of Group III dissolves completely in dilute HCl? **(b)** no precipitate is obtained on addition of NaOH and H_2O_2 (or Na_2O_2)? **(c)** the precipitate of Group III is white? **(d)** the filtrate from the H_2O_2 (or Na_2O_2) treatment is colorless? **(e)** the filtrate containing Group III is colorless?

4. If a test for iron is obtained in Procedure 10, how could you determine whether iron in the original sample was Fe^{2+} or Fe^{3+}?

5. Why use HNO_3 rather than HCl in testing for manganese?

6. If the precipitate formed on addition of NaOH and H_2O_2 (or Na_2O_2) is brick-red, what conclusions might be drawn?

7. What difficulties would be encountered if **(a)** chromium were incompletely oxidized by H_2O_2 (or Na_2O_2)? **(b)** the solution to which $BaCl_2$ is added is too acid? **(c)** the filtrate used in testing for Zn^{2+} is too acid?

8. Give the formula for a reagent that will **(a)** dissolve $Zn(OH)_2$ but not $Al(OH)_3$. **(b)** dissolve ZnS but not S. **(c)** form a precipitate with Al^{3+} but not Zn^{2+}. **(d)** form a precipitate with NaOH but not NH_3. **(e)** form a precipitate with NH_3 but not NaOH. **(f)** dissolve $Ni(OH)_2$ but not $Fe(OH)_3$. **(g)** dissolve $Al(OH)_3$ but not $Fe(OH)_3$. **(h)** form a precipitate with Co^{2+} but not Ca^{2+}.

Groups IV–V

1. Explain the separation of Ba^{2+} and Sr^{2+} from Ca^{2+}.

2. Why must a test for the ammonium ion be made on the original unknown?

3. Write the equation to show the action of a strong base on an ammonium salt.

4. Give the colors of the flame tests for Ca^{2+}, Ba^{2+}, Sr^{2+}, K^+, and Na^+.

5. Explain how a cobalt glass assists in the flame test for K^+.

6. How would you differentiate the following three white salts? **(a)** $MgCl_2$ **(b)** NaCl **(c)** KCl

7. Why is it necessary to sublime ammonium salts before testing for K^+?

8. How would you distinguish between **(a)** NaCl and NH_4Cl (solids)? **(b)** $MgCl_2$ and NH_4Cl (solids)? **(c)** $MgCl_2$ and KCl (solids)? **(d)** Mg^{2+} and Na^+ (solutions)?

9. Complete an analysis diagram for the separation of and ultimate testing for a solution containing only Fe^{3+}, Al^{3+}, Sr^{2+}, and Mg^{2+}.

All groups

1. From the following water-soluble salts—$CaCl_2$, K_3PO_4, $AgNO_3$, $CuSO_4$, $NaNO_3$—select three pairs that, when brought together in solution, will form a precipitate.

(a) ______________________

and ______________________

(b) ______________________

and ______________________

(c) ______________________

and ______________________

2. Give the formulas for two complex metallic ions of each of the following. **(a)** ammonia **(b)** hydroxide **(c)** sulfide **(d)** halide **(e)** cyanide

3. Indicate a reagent with which you could distinguish between the following solids. Record reactions for each.

Solids	Reagent	Results
(a) Hg_2Cl_2 and $HgCl_2$	__________	__________

(b) CuS and HgS ____________ ____________

(c) HgS and As_2S_5 . ____________ ____________

4. Show by equations how a single reagent might be used to separate the following pairs of ions. **(a)** Al^{3+} and Fe^{3+} **(b)** Cu^{2+} and Pb^{2+} **(c)** Hg^{2+} and Zn^{2+} **(d)** Cr^{3+} and Fe^{3+} **(e)** Ca^{2+} and Mg^{2+} **(f)** Ag^{+} and Hg^{2+} **(g)** Hg_2^{2+} and Hg^{2+} **(h)** Zn^{2+} and Al^{3+} **(i)** Ba^{2+} and Sr^{2+} **(j)** Cd^{2+} and As^{3+} **(k)** Sn^{2+} and Cu^{2+} **(l)** Bi^{3+} and Ni^{2+}

Complex Ions

1. From the following list of cations, select those that form amphoteric hydroxides. **(a)** Cd^{2+} **(b)** Cu^{2+} **(c)** Bi^{3+} **(d)** Sn^{2+} **(e)** Al^{3+} **(f)** Ni^{2+} **(g)** Cr^{3+} **(h)** Zn^{2+} **(i)** Mg^{2+}

Ans. **(d), (e), (g), (h)**

2. From the following list select those cations that form complex ions with NH_3. **(a)** Cu^{2+} **(b)** Hg^{2+} **(c)** Cd^{2+} **(d)** Bi^{3+} **(e)** Sn^{2+} **(f)** Cr^{3+} **(g)** Ni^{2+} **(h)** Zn^{2+} **(i)** Al^{3+} **(j)** Fe^{3+} **(k)** Ag^{+}

Ans. **(a), (c), (g), (h), (k)**

3. Show by equations a reagent that will dissolve one but not both of the following pairs (all solids). **(a)** $Al(OH)_3$ and $Fe(OH)_3$ **(b)** HgS and CuS **(c)** $PbCl_2$ and AgCl **(d)** $Zn(OH)_2$ and $Al(OH)_3$ **(e)** Na_2SO_4 and $PbSO_4$ **(f)** SnS and HgS **(g)** As_2S_3 and CdS **(h)** Sn and Cu **(i)** Sb and Ni **(j)** AgCl and AgI

4. Identify the following ions from the diagram shown below: A^{+}, B^{3+}, C^{2+}, D^{3+}, and E^{3+}.

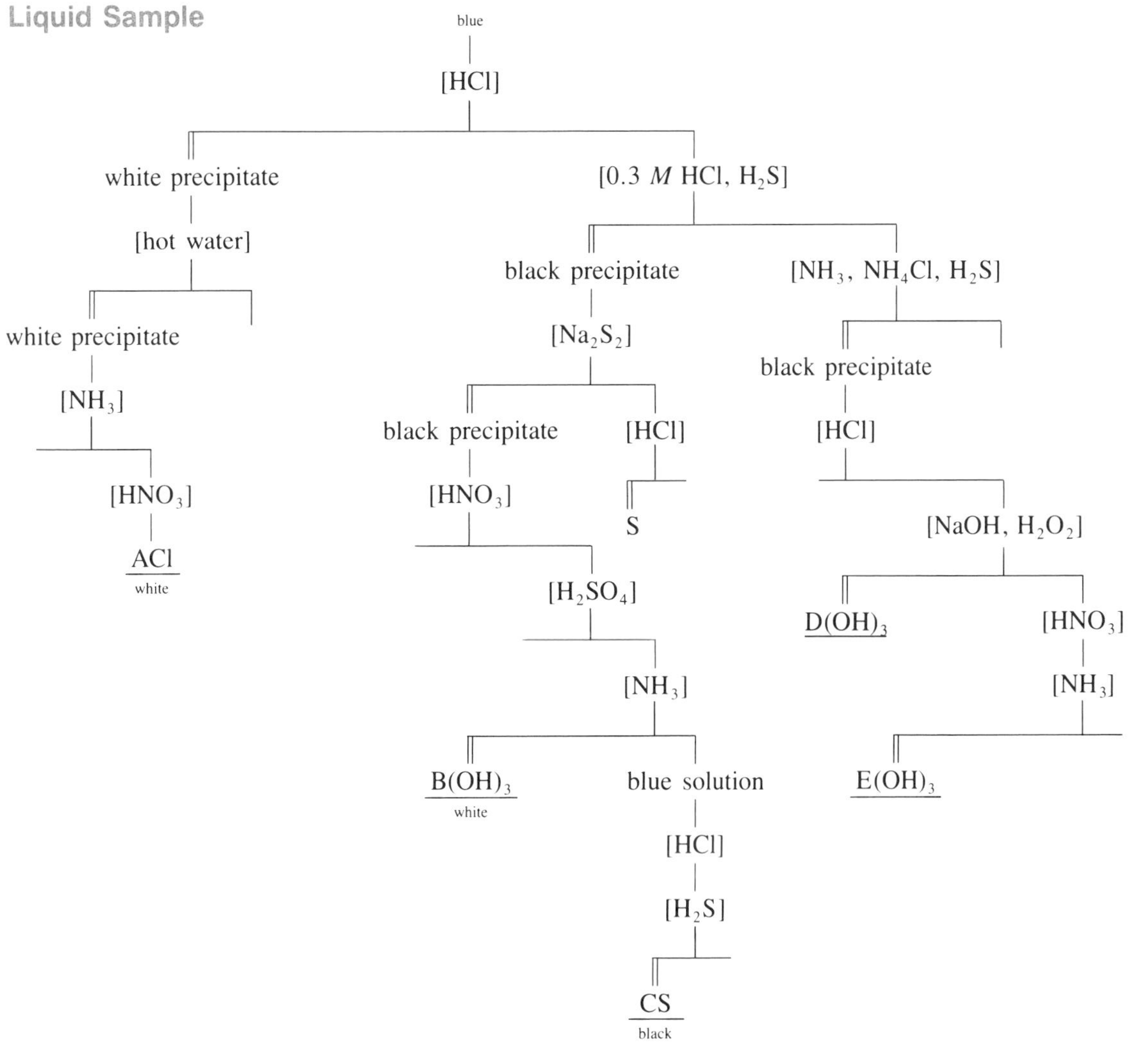

Alloy Analysis

1. Write equations for the reactions of concentrated HNO_3 with the following metals. **(a)** Sn **(b)** Sb **(c)** Fe **(d)** Cu **(e)** Ag

2. How would you distinguish chemically between the following metals? **(a)** Cu and Sn **(b)** Pb and Bi **(c)** Fe and Ag **(d)** Cr and Cu **(e)** Mg and Ag

3. Show diagrammatically the separation and identification of metals in alloys of: **(a)** Sn, Pb **(b)** Hg, Cu, Fe, Mg **(c)** Al, Cu, Sb

Analysis of Salts and Mixtures

1. A sodium salt analyzed for anions gave the following results. **(a)** Solid + concentrated H_2SO_4 gave a colorless gas with a sharp odor. **(b)** Aqueous solution gave no precipitate with $BaCl_2$ and NH_3, but gave a white precipitate with $AgNO_3$ and HNO_3.

Therefore, the salt is Na_______________.

2. A sodium salt analyzed for anions gave the following results. **(a)** Solid + concentrated H_2SO_4 produced no visible change. **(b)** Aqueous solution + $BaCl_2$, NH_3 gave a white precipitate soluble in HCl. **(c)** Solid + H_2SO_4 + methyl alcohol produced combustible gas that burned with a green flame.

Therefore, the salt is Na_______________.

3. The sulfide of a metal is insoluble in H_2O but soluble in cold dilute HCl. Its hydroxide is insoluble in H_2O but soluble in both NaOH and NH_3. What is the metal?

4. An unknown was found to be completely soluble in H_2O. Treatment of separate portions of the aqueous solution with Na_2HPO_4, $BaCl_2$, and $AgNO_3$ gave no precipitate. What cations and anions may be present?

5. An aqueous solution of a water-soluble white salt gives a precipitate with $(NH_4)_2CO_3$ in ammoniacal solution; it also gives a white precipitate with $AgNO_3$ and HNO_3. Which of the following salts is a possibility? **(a)** Hg_2SO_4 **(b)** NH_4Cl **(c)** $BaCl_2$ **(d)** $Al(NO_3)_3$ **(e)** $PbSO_4$

6. A salt is insoluble in water but dissolves in dilute HCl with effervescence of a colorless, odorless gas. A 0.3 *M* HCl solution of the salt gives a black precipitate with H_2S. Which of the following salts are possibilities? **(a)** $AlPO_4$ **(b)** NaBr **(c)** $KHCO_3$ **(d)** $SnCl_2$ **(e)** $PbCO_3$

7. A white solid is completely soluble in cold H_2O. Its aqueous solution is neutral to litmus. No change takes place when the solid is warmed with concentrated H_2SO_4, and there is no change when the aqueous solution is tested with H_2SO_4 and $FeSO_4$. Which of the following are possibilities? **(a)** Na_2SO_4 **(b)** NiS **(c)** $PbCrO_4$ **(d)** $(NH_4)_2CO_3$ **(e)** KNO_3

8. An unknown mixture of salts is completely soluble in H_2O and contains CO_3^{2-} and NO_3^- as the only anions. What cations may be present?

9. By means of a single reagent, how would you distinguish between the following pairs? Tell what happens in each case. **(a)** CdS and HgS (solids) **(b)** Sn^{2+} and Hg^{2+} (solutions) **(c)** Hg_2^{2+} and Hg^{2+} (solutions) **(d)** HCl and $HC_2H_3O_2$ (solutions) **(e)** Bi^{3+} and Cu^{2+} (solutions)

10. Tell how you would distinguish chemically between the following pairs of substances. In each case tell what happens to each substance.

Solutions	Solids
(a) $AgNO_3$ and $Zn(NO_3)_2$	**(a)** Hg_2Cl_2 and $HgCl_2$
(b) $Hg_2(NO_3)_2$ and $Hg(NO_3)_2$	**(b)** AgCl and $PbCl_2$
(c) $Pb(C_2H_3O_2)_2$ and $CuSO_4$	**(c)** Hg_2Cl_2 and $PbCl_2$
(d) $AgCl \cdot 2\ NH_3$ and $AgNO_3$	**(d)** AgCl and $BaCl_2$
(e) H_2SO_4 and HNO_3	**(e)** $BiCl_3$ and $ZnCl_2$
(f) HCl and HNO_3	**(f)** HgS and Bi_2S_3
(g) $HC_2H_3O_2$ and HCl	**(g)** PbS and SnS
(h) $HgCl_2$ and $SnCl_2$	**(h)** As_2S_5 and SnS_2
(i) $Pb(NO_3)_2$ and $CuSO_4$	**(i)** CuS and PbS
(j) $SnCl_2$ and $SnCl_4$	**(j)** CuS and Sb_2S_5
(k) $HgCl_2$ and $ZnCl_2$	**(k)** $Bi(OH)_3$ and $Pb(OH)_2$
(l) $(NH_4)_3AsS_4$ and $FeCl_3$	**(l)** FeS and NiS
(m) $CuSO_4$ and $Bi_2(SO_4)_3$	**(m)** FeS and MnO_2
(n) $Pb(NO_3)_2$ and $Bi(NO_3)_3$	**(n)** NiS and $Ni(OH)_2$
(o) $SbCl_3$ and $BiCl_3$	**(o)** $Zn(OH)_2$ and ZnS
(p) $PbCl_2$ and $BaCl_2$	**(p)** $AlCl_3$ and $Al(OH)_3$
(q) H_2S and HCl	**(q)** Na_2CrO_4 and $BaCrO_4$
(r) HCl and H_3AsO_4	**(r)** MnS and NiS
(s) $PbCl_2$ and $BiCl_3$	**(s)** NaCl and NH_4Cl
(t) $BiCl_3$ and $AlCl_3$	**(t)** $CaCO_3$ and Na_2SO_4
(u) $AlCl_3$ and $ZnCl_2$	**(u)** Ag and Zn
(v) $CrCl_3$ and $NiCl_2$	**(v)** Sn and Ni
(w) $ZnCl_2$ and $NiCl_2$	**(w)** Sn and Sb
(x) $MnCl_2$ and $ZnCl_2$	**(x)** Pb and Sb
(y) $FeCl_3$ and $FeCl_2$	**(y)** $BaCO_3$ and $BaSO_4$

11. A white salt (single) is completely soluble in cold H_2O and is also soluble in excess NaOH solution or excess NH_3. No precipitate is obtained with H_2S in 0.3 *M* HCl solution, but, when the solution is made basic with NH_3 and H_2S added, a precipitate is obtained. The original sample treated with concentrated H_2SO_4 yields a brown gas. An aqueous solution of the salt gives a pale yellow precipitate with $AgNO_3$ in HNO_3 solution. What is the salt?

12. An unknown mixture of salts is completely soluble in cold H_2O. The colorless solution gives a brick-red flame test. When the solution is treated with $AgNO_3$, no precipitate is obtained. What anions may be present?

13. Which of the following pairs of salts in aqueous solution would give no precipitate upon mixing?
(a) $BaCl_2$—$AgNO_3$ **(b)** Hg_2SO_4—Na_2S
(c) NH_4Cl—$CdSO_4$ **(d)** KBr—$Pb(NO_3)_2$
(e) Na_3AsO_4—$Ca(C_2H_3O_2)_2$

14. A copper salt is insoluble in H_2O. A prepared solution (made by treating the solid with Na_2CO_3, filtering, and boiling the filtrate with dilute HNO_3) gave no precipitate with NH_3 and $BaCl_2$. What anion(s) may be present?

APPENDIX

Equipment

Individual equipment

1 beaker, 400 mL	1 iron ring, 4 inches
1 beaker, 150 mL	1 triangle
2 beakers, 50 mL	1 ring stand
1 test tube brush	1 test tube rack (for 3-inch test tubes)
1 box matches	1 reagent block
1 Bunsen burner (micro) and tubing	14–15 mL dropping bottles
1 burette clamp	6 medicine droppers
1 test tube holder	18 3-inch test tubes
1 crucible, No. 0	1 watch glass
1 graduated cylinder, 10 mL	1 wire gauze
1 evaporating dish, 3 inches	1 wing top
1 file	1 Florence flask, 250 mL
1 pair forceps	2 vials litmus paper
5 pieces 6 mm glass tubing, 18 inches	filter paper
6 pieces glass rod, 3 mm	3 towels
1 spatula (metal)	1 apron
	1 platinum wire (or nichrome wire)
	1 sponge
	1 pair safety glasses or goggles

Centrifuges, 1 for each 8–10 students

Set of reagents, liquid and solid (see below)

Sideshelf Reagents for Semimicro Analysis

(Liquids in 100 mL dropping bottles)

Reagent	Directions for making solutions
Acetic acid, 6 *M*	350 mL glacial acetic acid and 650 mL H_2O
Alcohol, absolute, ethyl or Isopropyl alcohol, dry	100%
Aluminon	0.1% solution of ammonium aurin tricarboxylate

(Liquids in 100 mL dropping bottles)

Reagent	Directions for making solutions
Ammonium acetate	250 g $NH_4C_2H_3O_2$ in 1 liter H_2O
Ammonium carbonate	250 g $(NH_4)_2CO_3$ in 1 liter 6 *M* NH_3
Ammonium chloride	100 g NH_4Cl in 1 liter H_2O
Ammonia, 6 *M*	400 mL concentrated NH_3 and 600 mL H_2O
Ammonia, 15 *M*	Concentrated reagent, as received
Ammonium molybdate, 0.5 *M*	Dissolve 50 g MoO_3 in a mixture of 135 mL H_2O and 75 mL concentrated NH_3. Add this solution slowly and with stirring to a mixture of 245 mL concentrated HNO_3 and 575 mL H_2O. Makes 1 liter.
Ammonium oxalate	40 g $(NH_4)_2C_2O_4$ in 1 liter H_2O
Ammonium sulfate	132 g $(NH_4)_2SO_4$ in 1 liter H_2O
Arsenazo	0.001 *M* or 0.65 g/liter of $C_{16}H_{10}AsN_2Na_3O_{11}S_2 \cdot 2\ H_2O$
Barium chloride	120 g $BaCl_2 \cdot 2\ H_2O$ in 1 liter H_2O
Barium hydroxide	Saturated solution (calcium hydroxide may be used)
Chlorine water	Saturate water by bubbling chlorine gas through it
Dimethyl glyoxime	1 g in 100 mL 95% ethyl alcohol
Dithizone test paper	Soak filter paper in a solution of 1 g dithizone in 25 mL chloroform and dry
Ether	
Ethyl alcohol	95%
Hydrochloric acid, 1 *M*	Dilute 80 mL 12 *M* HCl to 1 liter
Hydrochloric acid, 6 *M*	Dilute concentrated HCl, 12 *M*, 1 to 1
Hydrochloric acid, 12 *M*	Concentrated reagent, as received
Hydrogen peroxide	3% (Check for freshness or stability)
Iodine–potassium iodide mixture	Dissolve 168 g KI in 1 liter of H_2O and add 25 g iodine
Lanthanum nitrate	5% aqueous solution
Lead acetate	190 g $Pb(C_2H_3O_2)_2 \cdot 3\ H_2O$ in 1 liter H_2O
Magnesia mixture	Dissolve 130 g $Mg(NO_3)_2 \cdot 6\ H_2O$ and 240 g NH_4NO_3 in H_2O, add 35 mL 6 *M* NH_3 and dilute to 1 liter
Mercuric chloride	27 g $HgCl_2$ in 1 liter H_2O
Methyl alcohol	
Nitric acid, 16 *M*	Concentrated reagent, as received
Nitric acid, 6 *M*	Dilute 375 mL concentrated HNO_3 to 1 liter
Potassium chromate	97 g K_2CrO_4 in 1 liter H_2O
Potassium cyanide	65 g KCN in 1 liter H_2O (KEEP IN STOREROOM)
Potassium ferricyanide	100 g $K_3Fe(CN)_6$ in 1 liter H_2O
Potassium ferrocyanide	105 g $K_4Fe(CN)_6$ in 1 liter H_2O
Silver nitrate	17 g $AgNO_3$ in 1 liter H_2O
Silver sulfate	5 g Ag_2SO_4 in 1 liter H_2O
Sodium carbonate	Saturated solution
Sodium acid phosphate	107 g Na_2HPO_4 in 1 liter H_2O
Sodium hexanitrocobaltate(III)	Dissolve 25 g $NaNO_2$ in a mixture of 50 mL H_2O and 15 mL 6 *M* acetic acid. Add 2.5 g $Co(NO_3)_2 \cdot 6\ H_2O$. Allow to stand for 24 hours, then filter and dilute to 100 mL (Fresh solution essential. Solid reagent may be stored.)
Sodium hydroxide, 5%	Dissolve 50 g NaOH in 1 liter H_2O
Sodium hypochlorite	Commercial bleach, e.g., "Purex"
Sodium polysulfide	Dissolve 480 g $Na_2S \cdot 9\ H_2O$ and 40 g NaOH in H_2O. Dilute to 1 liter. Dissolve 10 g flowers of sulfur in the solution.
Sulfuric acid, 18 *M*	Concentrated reagent, as received
Sulfuric acid, 3 *M*	Dilute 165 mL concentrated H_2SO_4 to 1 liter

(Liquids in 100 mL dropping bottles)

Reagent	Directions for making solutions
Thioacetamide	8% aqueous solution
Tin(II) chloride	Dissolve 100 g $SnCl_2 \cdot 2\ H_2O$ in 200 mL concentrated HCl and dilute to 1 liter. Add a few pieces of metallic tin.
Zinc uranyl acetate	Dissolve 10 g $UO_2(C_2H_3O_2)_2 \cdot 2\ H_2O$ in 6 mL 30% acetic acid with heat, if necessary, and dilute to 50 mL. Mix 30 g $Zn(C_2H_3O_2)_2 \cdot 3\ H_2O$ and 3 mL 30% acetic acid, and dilute to 50 mL. Mix solutions, add pinch of NaCl. After 12 hours, filter.

Cation Test Solutions

Test solutions of all cations should be available. The concentration of these test solutions should be approximately 10 mg of the ion per milliliter of solution (about 0.5 mg per drop).

Ion	Directions for making solutions
Ag^+	17 g $AgNO_3$ in 1 liter H_2O
Hg_2^{2+}	Dissolve 14 g $Hg_2(NO_3)_2 \cdot 2\ H_2O$ in 100 mL 6 *M* HNO_3 and dilute to 1 liter
Pb^{2+}	16 g $Pb(NO_3)_2$ in 1 liter H_2O
Cu^{2+}	38 g $Cu(NO_3)_2 \cdot 3\ H_2O$ in 1 liter H_2O
Hg^{2+}	17 g $Hg(NO_3)_2 \cdot \frac{1}{2}\ H_2O$ in 1 liter H_2O
Bi^{3+}	Dissolve 19 g $BiCl_3$ in 1 liter of 2 *M* HCl
Cd^{2+}	28 g $Cd(NO_3)_2 \cdot 4\ H_2O$ in 1 liter H_2O
As^{3+}	Dissolve 14 g As_2O_3 in 1 liter 6 *M* HCl
Sn^{2+}	19 g $SnCl_2 \cdot 2\ H_2O$ in 1 liter 6 *M* HCl
Sn^{4+}	30 g $SnCl_4 \cdot 5\ H_2O$ in 1 liter 6 *M* HCl
Sb^{3+}	19 g $SbCl_3$ in 1 liter 3 *M* HCl
Co^{2+}	48 g $CoCl_2 \cdot 6\ H_2O$ in 1 liter H_2O
Ni^{2+}	48 g $Ni(NO_3)_2 \cdot 6\ H_2O$ in 1 liter H_2O
Mn^{2+}	54 g $Mn(NO_3)_2 \cdot 6\ H_2O$ in 1 liter H_2O
Zn^{2+}	46 g $Zn(NO_3)_2 \cdot 6\ H_2O$ in 1 liter H_2O
Fe^{3+}	50 g $FeCl_3 \cdot 6\ H_2O$ in 10 mL concentrated HCl and 1 liter H_2O
Al^{3+}	140 g $Al(NO_3)_3 \cdot 9\ H_2O$ in 1 liter H_2O
Cr^{3+}	51 g $CrCl_3 \cdot 6\ H_2O$ in 1 liter H_2O
Ba^{2+}	20 g $Ba(NO_3)_2$ in 1 liter H_2O
Sr^{2+}	33 g $Sr(NO_3)_2 \cdot 4\ H_2O$ in 1 liter H_2O
Ca^{2+}	59 g $Ca(NO_3)_2 \cdot 4\ H_2O$ in 1 liter H_2O
Mg^{2+}	105 g $Mg(NO_3)_2 \cdot 6\ H_2O$ in 1 liter H_2O
K^+	25 g KNO_3 in 1 liter H_2O
Na^+	25 g NaCl in 1 liter H_2O
NH_4^+	30 g NH_4Cl in 1 liter H_2O
Li^+	60 g LiCl in 1 liter H_2O

Unknown Solutions

Unknown solutions should be made up to contain approximately 10 mg of each ion per milliliter of solution (see above).

Solids

Ammonium chloride
Ammonium thiocyanate
Iron(II) sulfate
Magnesium (granular or ribbon)
Methyl violet test paper
Oxalic acid
Potassium chlorate
Sodium acetate
Sodium arsenate
Sodium bismuthate
Sodium bromide
Sodium carbonate
Sodium chloride
Sodium chromate
Sodium iodide
Sodium nitrate
Sodium nitrite
Sodium peroxide
Sodium phosphate (Na_2HPO_4)
Sodium sulfate
Sodium sulfite
Sodium tetraborate
Zinc (granular)
Zinc sulfide or calcium sulfide

INDEX

Acetate, test for, 27, 30
Acute poisons, 1–2
Al^{3+}, 61
Alkali group, 68–72
  analytical procedures for, 70–72
  chemistry for, 69–70
  diagram of, 68
  separation of barium group and, 63–67
  unknown solution analysis, 72
Alloys, 75–78
  common alloys list, 75
  problems for, 96
  unknown alloy analysis, 76–78
Aluminum–nickel–iron group, 53–62
  chemistry of, 53–57
  diagram of, 54
  unknown solution analysis, 62
Aluminum subgroup, 60–61
Ammonium salts, 71, 72
Anions, 3, 19
  group tests for, 22–23
  mixtures, analysis of, 32
  problems in analysis, 93
  procedures for specific tests, 28–30
  in rocks, 84
  of salts, 80
  solubility group tests, 22–23
  specific tests for, 24–32
  sulfuric acid and, 20–21
  unknowns, analysis of, 31
Antimony salts, 47
Appendix, 98–101
Arsenate, test for, 24–25, 28
Arsenic subgroups, 43–52

$BaCrO_4$, 61
Barium group, 63–67
  analytical procedures for, 64–67
  diagram for, 64
Bismuth hydroxide, 47
Borate, test for, 25, 28
Borax bead, 14
Bromide, test for, 26–27, 29
Bronze alloy, 76

Carbonate, test for, 26, 29
Cations, 3, 19, 33–72
  aluminum–nickel–iron group, 53–62
  appendix of test solutions, 100
  barium group, 63–67
  copper–arsenic group, 43–52
  group separations, 34
  method for analysis, 35–36
  preliminary experiment, 37–38
  problems in analysis, 93–95
  of salts, 80
  silver group, 39–42
Centrifuge, 2
  defined, 7
  laboratory technique experiment, 15–16
  safety with, 2
  use of, 8, 9
Chemical solubility, 3
Chloride, test for, 26–27, 29
Chromate, test for, 25–26, 28
Chromium chemistry, 60–61
Chromium salts, 2
Chronic poisons, 2
Cleanliness, 9
Colloidal precipitates, 7
Colored ions, 79
Commercial substances, 83–84
Complex ions, 95
Contact lenses, 1
Copper–arsenic group, 43–52
  analytical procedures for, 47–52
  diagram of, 44
  unknown, analysis of, 52
Copper subgroups, 43, 45, 49–51

Decantate, 9
Decantations, 7
Diagrams
  cation group separations, 34
  group I: the silver group, 39
  group II: the copper–arsenic group, 44
  group III: the aluminum–nickel–iron group, 54
  group IV: the barium group, 64
  group V: the alkali group, 68
  liquid sample, 95

Equations, 92
Equipment, 8–9
  appendix information, 98
  see also Centrifuge
Evaporations technique, 9
Eye protection, 1

Ferrous alloys, 76, 77
Filtrate, 7
Filtration, 7
Fire hazards, 1
Flame test, 13
  in alkali group, 71
Fume hoods, 2

Glasses, requirement for, 1

Halide group anions, 22
Halogenated organic solvents, 2
Halogens, separation of, 31
HF, safety rules for, 2
H_2O_2, 57
H_2S gas, 2
  thioacetamide as source of, 15–16
Hydrochloric acid, 35

Ignition, 7
Industrial-process separations, 3
Industrial substances, 83–84
Inorganic substance analysis, 83–84
Iodide, test for, 26–27, 29
Ionization constants, 88–89
Ions
  colored ions, 79
  complex ions, 95
Iron subgroup, 56–57, 59–60

K^+, 71

Laboratory techniques
  with centrifuge, 15–16
  problems involving, 92–93

Lead sulfate, 47
Liquid sample, 95

Major ingredients in alloys, 76
Methyl violet, 47
Mg^{2+}, 71
Minerals, analysis of, 84
Minor ingredients in alloys, 76
Mixtures. *See* Salts and mixtures

Na^+, 71
NaOH, 57
NH_3, 57
NH_4^+, 71
NH_4Cl, 57
Nichrome wire, 13
Nickel subgroup, 56, 59
Nitrate
  anions group, 23
  test for, 27, 29
Nitric acid, 76
Nitrite, test for, 27, 30
Nonferrous alloys, 76, 77

Outlines for cation analysis, 36

Pasteur pipets, 13
*p*H, problems involving, 89
Pharmaceuticals, 83–84
Phosphate, test for, 25, 28
Physical solubility, 3
Pipets, 13
Platinum wire, 13
*p*OH, problems involving, 89
Poisons, 1–2
  acute poisons, 1–2
  chronic poisons, 2
Potassium, 71, 72
Precipitates
  for alkali group, 71
  for barium group, 65
  for copper and arsenic subgroups, 48–52
  for silver group, 41
  for subgroups of group III, 58–60
Precipitation, 7
  of aluminum–nickel–iron group, 53, 55, 57–58
  of arsenic subgroups, 43
  of copper subgroup, 43
  handling of, 9
Preliminary experiments
  on aluminum–nickel–iron group, 57
  for anion separation, 27
  for cation analysis, 37–38
  for copper–arsenic group, 47
  for silver group, 40
Prussian blue, 3

Reagents, 7
  sideshelf reagents, 98–100
Rocks, analysis of, 84

Safety rules, 1–2
Salts and mixtures, 79–82
  problems for analysis, 96–97
  sample data sheet for, 81–82
  unknowns, analysis of, 81
Schedule of laboratory work, 11
Sideshelf reagents, 98–100
Silicates, 84
Siliceous rocks, 84
Silver group, 39–42
  analytical procedure for, 40–42
Sodium salts, 19
Solids, appendix material, 101
Solubility
  anions, tests of, 22–23
  problems involving, 90–91
  rules for, 9–10
Solutions
  for alkali group, 70
  for barium group, 65
  for copper and arsenic subgroups, 48–51
  problems pertaining to, 87–88
  for silver group, 41
  for subgroups of group III, 57–60
Solvents for salts, 79–80
Spatulas, 12–13
Squeeze bottle, 12
Stirring rods, 12–13
Sulfate
  anions, 22
  test for, 24, 28
Sulfides
  in copper–arsenic group, 47
  test for, 26, 29
Sulfite, test for, 26, 28
Sulfuric acid, 20–21
Supernatant liquid, 9

Tartrate, 30–31
Tests, 7
Thioacetamide, 2
  as H_2S source, 15–16
Trace amounts, 76
Turbidity, 7

Wash bottle, 12
Washing, 7
  techniques, 9

Zinc, 61
  NaOH and, 57
ZnS, 61